VIRUS AND PLANT DISEASES

VIRUS
AND
PLANT DISEASES

By

Dr. Shubhrata R. Mishra

Department of Botany
Vikram University
Ujjain (M.P)

DISCOVERY PUBLISHING HOUSE
NEW DELHI-110002

First Published-2004
Reprinted: 2013
ISBN 81-7141-779-5

Published by

DISCOVERY PUBLISHING HOUSE
4831/24, Ansari Road, Prahlad Street,
Darya Ganj, New Delhi-110002 (India)
Phone: 23279245 • Fax: 91-11-23253475
E-mail:dphtemp@indiatimes.com

Printed at: Dynamic printers, Delhi

Preface

The scope of virology is expanding so rapidly that it is impossible to present all of it in a book which a student new to the field can cover in a single course. We have, therefore, tried to present selected portions of virology in sufficient detail that the student can understand them through reading the book. The primary aim of this text book is to present a succinet account of the essential features of the virus in a form suitable for students. Efforts have been made to include only sound fundamental material to give the beginner a solid foundation for more advanced work on the subject. Emphasis is placed on the use of chemistry for a clearer understanding of the composition of viruses and the reactions they produce.

The book uses up-to-date examples to show how precise hypotheses may be formulated and tested experimentally. Special efforts have been made to explain ideas in non mathematical terms. The primary aim throughout has been clarity, simplicity and the high standard. The book will definitely prove to be a boon to teachers, students and research workers in the related field.

The author is indebted to her colleagues for their aid in reading and checking the proofs, and to all who have offered valuable suggestions and criticisms during the preparation of the manuscript.

Any errors or omissions are entirely unintentional. She alone accepts full responsibility for any defects that may be inheret in the plan and scope of the book and for any errors which may have escaped detection.

The author expresses her gratitute to Mr. Wasan and staff of M/s Discovery Publishing House for their whole hearted co-operation in the publication of this book.

Author

CONTENTS

1

INTRODUCTION

ORIGIN OF VIRUS

We have known for a long time that viruses can have disastrous effects on plants and animals. Viruses are not plants, animals, or bacteria, but they are the quintessential parasites of the living kingdoms. Although they may seem like living organisms because of their prodigious reproductive abilities, viruses are not living organisms in the strict sense of the word. The word "virus" had long been used for any slimy liquid, poison, venom or infectious matter. The original meaning of virus can still be detected in the commonly used adjective "virulent", meaning extremely piosinous, venomous or malignant. Without a host cell, viruses cannot carry out their life-sustaining functions. They parasitize the cell for basic building materials, such as amino acids, nucleotides, and lipids (fats). Although viruses have been speculated as being a form of protolife, their inability to survive without living organisms makes it highly unlikely that they preceded cellular life during the Earth's early evolution. Some scientists speculate that viruses started as rogue segments of genetic code that adapted to a parasitic existence. The probably multiple origins of viruses are lost in a sea of conjecture and speculation, which results mostly from their nature. No one has ever detected a fossil virus as a particle. They are too small and probably too fragile to have withstood the kinds of processes that led to fossilisation, or even to preservation of short stretches of nucleic acid sequences in leaf tissues or insects in amber. Now a days, the new science of virus molecular systematics is shedding a great deal of light on the distant relationships and presumed origins of many important groups of viruses. This is as a result of the sequencing of all or part of the genomes of representatives of many of the known varieties of viruses including the largest (pox- and herpesviruses) and the smallest (gemini and other ssDNA viruses). If viral genomes are compared with each other and with cellular sequences, presumed patterns of evolution/ divergence of the genomes can be reconstructed.

Evolution of Virus

The evolution of viruses in nature is performed in two levels. Microevolution means smaller changes of genome by means of the substitution, addition or deletion of nucleotides and leads to the creation of strains and other forms within the species. Macroevolution leads to sudden major changes by duplication of existing genes, by recombination with related viruses or by the acquisition of a gene from the host or an unrelated virus, which lead to the creation of new species. By these ways of evolutionary development, viruses, as spiritless wonders of nature, have brought to the perfection their forms, architecture and replication mechanisms by using different and always efficient ways and means for the realization of the purpose of their existence - to produce infectious progeny. One major assumption has to be made that viruses co evolve with their hosts, like any good parasite. The studies on some viruses, such as papillomaviruses, endogenous retrovirus-like sequences in animal genomes, and herpesviruses, have justified their co evolve concept. The divergence of primates and of birds related to chickens has been traced by comparing the types and sequences of retroviral-derived sequences in their genomes. It has also been repeatedly shown that the closest relatives of human papillomavirus types infecting particular tissue types (eg: cutaneous wart types, genital mucosal types) are those viruses infecting similar tissue types in other primates, indicating that these tissue preferences were well established before the divergence of humanoid apes from the primate line.

It is quite useful to consider the timeline of evolution of life from its beginnings in water, as well as the timeline of colonisation of dry land by organisms and terrestrial organisms. This process went much as follows

- 4000 Myr: life originated in the sea/shallow ponds
- 3500 Myr: bacteria emerged from the seas to colonise the land

- 1000 Myr: plants/fungi slowly colonised inland of coastal margins
- 700 Myr: insects/arthropods crawled out to feed on plants
- 350 Myr: vertebrates adapted to air breathing/survival on land

Table 1.1. Differences between viruses and cells

Characteristic	*Virus*	*Cell*
Covering	protein capsid = simple, only protein	complex membrane - specialized proteins associated with other macromolecules
Nucleic acid	simple - only content	various and complex - integrated with other components
Genome	few genes - ±250 - coding for few enzymes; enzyme complement incomplete for reproduction; no machinery for energy utilization	many genes - enormous enzyme enzyme complement even in simple cell; sufficient for self-reproduction
Replication Reproduction	virions do not grow in number requires virion break-down	cell do grow in number no break-down required-integrity maintained at all times
Origin	virion never arises from another virion directly	cell arise from other cells
Division	virion produces many other virions in 1 step	cell undergoes division-1 cell-2 cells in 1 step
Dependency	virions need cells as hosts for life-cycle to occur - dependent	cells are self-maintaining - independent

Viruses of nearly all the major classes of organisms - animals, plants, fungi and bacteria/archaea - probably evolved with their hosts in the seas, given that most of the evolution of life on this planet has occurred there. This means that viruses also probably emerged from the waters with their different hosts, during the successive waves of colonisation of the terrestrial environment. Thus,

- Viruses of terrestrial bacteria probably derive from those of the original colonizers;
- Most viruses of land plants are probably evolved from those in the green algae that emerged +/- 1000 Myr ago;

- Most viruses of terrestrial arthropods derive from the first founders;
- Most viruses of terrestrial vertebrates derive from those that came ashore with the first air-breathers.

This would explain why viruses in the different broad types of hosts are generally so different: they have had aeons to adapt to each "life niche" since divergence from the respective common ancestor. Thus, bacteria/archaea and eukaryotes share no virus types, as they have been diverged so long. However, there are some marked similarities between viruses infecting plants and those infecting vertebrates. A complicating factor in the picture of viruses co-evolving with their hosts over millennia is the fact that viruses apparently can - and obviously do - make big jumps in hosts every now and then. It seems obvious, for example, that arthropods are almost certainly the original source for a number of virus families infecting insects and mammals - such as the Flaviviridae - and probably also of viruses infecting insects and other animals and plants - such as the Rhabdoviridae and Reoviridae. Picornaviruses of mammals are very similar structurally and genetically to a large number of small RNA viruses of insects and to at least two plant viruses, and - as the insect viruses are more diverse than the mammalian viruses - probably had their origin in some insect that adapted to feed on mammals (or plants) at some distant point in evolutionary time.

Iues as to how this happens are given by the fact that flock house virus - a small naked isometric ssRNA virus of insects - can replicate in plants when introduced into leaves, but only at the site of infection. This gives it a survival advantage over other similar viruses, which cannot replicate in plants, as these are now no longer a passive, but an active reservoir for the re-infection of insects. For example, leafhopper A virus and Rhopalosiphum padi [aphid] virus are both stable enough and are injected into plants at sufficient concentration by their hosts, to make the plants into "circulative non-propagative" vectors of the viruses. The addition of a movement function - a "module" peculiar to plant viruses - would make flock house virus into a fully-fledged plant virus as well; loss of ability to infect insects would then limit it to plants.

It is obvious that a number of such crossovers have occurred: there are reoviruses, which only infect plants; flaviviruses which only infect humans; picornaviruses which only infect animals or plants, and so on. An interesting "emerging virus" is tomato spotted wilt virus - genus Tospovirus, family Bunyaviridae - which is postulated to have relatively recently acquired the ability to infect plants as well as its thrips vector insect, as well as obtaining a movement function, and is now a major pathogen of a wide range of plant species.

Another corollary of considering cellular organismal evolution as a marker of virus evolution is that the terrestrial virus gene pool must constitute only a subset of the virus lineages currently present in the oceans: if all of the major lines of terrestrial organisms are descended from subsets of parallel-developing lineages in the oceans, then the terrestrial ancestral gene pool is smaller than the oceanic. Assuming virus diversity parallels host diversity, therefore the ancestral terrestrial virus gene pools are also smaller than their oceanic equivalents.

Phylogeny of Virus

The deep phylogenetic studies are showed that early virus evolution was almost certainly modular. Certain core modules that proved to be successful - like the retrovirus pol gene, and the picornavirus-like protease-Vpg-polymerase module - appear in a

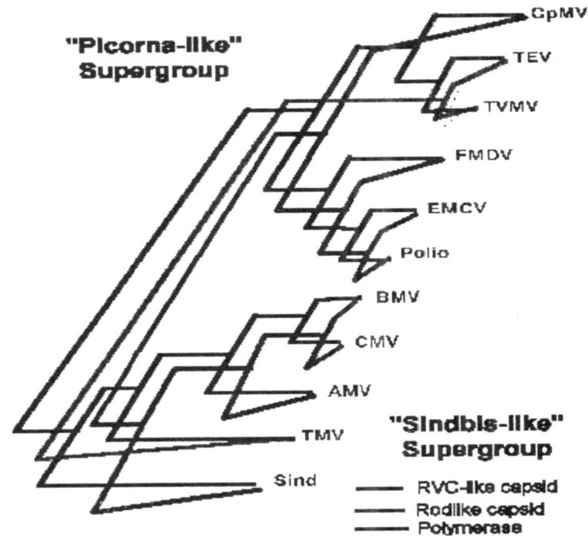

Fig. 1.1. Phylogeny of virus.

number of different contexts. Thus, certain animal viruses - like picornaviruses and alphaviruses - have relatives among plant viruses, which do not necessarily share the same morphology, number of genome components, or even genome organisation or number of genes.

For example, small spherical picornaviruses (ssRNA, 1 component, infect animals) are related to comoviruses (small spherical, 2 component, plant) and Potyviridae (filamentous, 1 or two genome components, plant), as part of a "Picorna-like Supergroup". Similarly, Sindbis (ssRNA, enveloped spherical, single RNA, animals) virus is related to Bromoviridae (naked spherical, 3 component, plant) and Tobamoviruses (naked rod, 1 component, plant), in a "Sindbis (or Alpha-) virus-like Supergroup". A diagram shows overlaid phylogenetic relationships of different components of some well-known viruses, illustrating how different components of a group of viruses may have different "gene trees". In this case, it shows how individual viruses in two different "supergroups" of viruses - defined in terms of polymerase affinities - have different gene trees when capsid proteins and polymerase relationships are taken into account. Thus, for ss(+)RNA viruses at least, going back beyond a certain level results in a severe blurring of perceived relationships, as well-defined families of viruses appear to share essential core components with other well-defined families / orders, while having nothing else in common.

It is very quickly apparent from sequence studies that there can have been no single origin of viruses as organisms. For instance, there is no obvious way one can relate viruses of the size and complexity of the Poxviridae [double-stranded linear DNA,130-375 kb, 150-300 genes] with viruses like the tobamoviruses [ss linear RNA, 6-7 kb, 4 genes], or either of these with the Geminiviridae [ss circular DNA, 2.7 - 5.4 kb, 3-7 genes]. Thus, there can be no simple "family tree" for viruses; rather, their evolutionary descent must resemble a number of scattered "bushes". Viruses as a class of organism must be therefore are considered to be polyphyletic in origins: that is, having a number of independent origins, almost certainly at different times, usually from cellular organisms.

History of Virus

There seem to be agreement that the history of viruses dates from the end of the 19th century. This is attributed to the

Dmitri Iwanowsk
(1864-(1920)

Martinus Beijernick
(1851-1931)

observations of Adolf Mayer in Germany, Dimitri Ivanofsky in Russia and Martinus Beijerinck in the Netherlands, all of whom were led to their discoveries during their investigations of diseased tobacco plants.In 1886, Adolf Mayer had demonstrated that when juice from tobacco plants infected with the mosaic disease was injected into healthy plants, it reproduced the mosaic disease. Boiling the juice destroyed the infectivity. Mayer thought that the causative agent was a bacterium. Inoculation of tobacco plants with a variety of bacteria, however, failed to produce the tobacco mosaic disease.

Ivanovski confirmed the observations of Mayer and also made another very important one. He presented his paper related to virology to the St. Petersburg Academy of Science on the 12th February 1892. He showed that extracts from diseased tobacco plants could transmit disease to other plants after passage through ceramic filters fine enough to retain the smallest agent known as bacteria. This agent was later called a virus.

After six years, Martinus Beijernick confirmed Iwanowski's results on tobacco mosaic virus. He discovered that the agent causing the disease could reproduce itself. Beijerinck was the first to make a connection between the filtrates and their infectious capability; he reasoned that some infectious agent, smaller than a bacterium must have been making the plants sick. It is notable that at this time, viruses as microbes were not known and the term "filterable disease agent" was used. He developed with the terme "contagium vivum fluidum" or "contagious living liquid" ('soluble living germ') as first the idea of the virus. Agents that pass through filters that retain bacteria came to be called ultrafilterable viruses, approrating the term virus from the Latin for "poison". The same year (1898), the German scientists Friedrich Loeffler (1852-1915) and Paul Frosch, both former students and assistants of Robert Koch (1843-1910), observed that a similar agent was responsible for foot-and-mouth disease. In spite of these findings, there was resistance to the idea that these mysterious agents might have anything to do with human diseases. These pioneering works on tabacco mosaic- and foot-and-mouth disease virus was followed by the identification of viruses associated with specific diseases in many other organisms. In 1900, the first human disease which known to be caused by a filterable agent was Yellow fever (a mosquito-transmitted disease), recognized by the US Army under the direction of Sir Walter Reed. Karl Landsteiner discovered poliovirus in 1909. Before the discovery of the infectious nature of the disease, the paralytic aspect was considered to be its characteristic feature, as documented by the denomination "infantile paralysis." Although its infectious nature was long hypothesized, Ivar Wickman was the first to clearly show the infectious nature of polio after an epidemic in Sweden in 1905. In 1911, Peyton Rous (1879-1970) was able to prove that some spontaneous chicken tumours, to all appearances classical neoplasms, are actually started off and driven by viruses (Rous sarcoma virus), which determine their forms as well. These findings led him to spend several years trying to get similar agents from mouse cancers; but, failing in this, he left off working with tumours in 1915, turning instead to the study of other problems in physiological pathology. Bacterial viruses were first described by Frederick Twort in 1915 and Felix d'Hérelle in 1917. D'Hérelle named them bacteriophages because of their ability to lyse bacteria on the surface of agar plates. The researchers noted the significance of plaque formation, whereby lysis of bacterial cells by phages would leave visible clearings on petri plates. Plaque formation is an important tool in "phage typing" or identification of specific bacterial strains. D'Herelle also recognized the significance of phage therapy. Humans could receive phages in order to treat specific bacterial infections. Research on phage therapy fell by the wayside as antibiotics became popular. However, as antibiotic resistance is a serious threat phage therapy is once again being considered as a realistic alternative to antibiotics. The electron microscope became available in the late 1930's and for the first time viruses could be visualized. Negative staining techniques were also developed as powerful tools for analyzing viral particles. As a science, the field of virology is a relatively young one. Once the structure of DNA and the nature of the genetic code were elucidated in the 1950's, rapid advances were made in understanding the molecular biology of viruses. Rapidly many scientists utilized these viruses as model systems to investigate many aspects of virology, including virus structure, genetics, and replication.

Occurrence of Virus

All viruses are parasitic in cells and cause a multitude of diseases in all forms of living organisms, from single-celled microorganisms to large plants and animals.

Host Range of Virus

- A virus can only infect and reproduce within certain host cells.
- Viruses identify their hosts by specific receptor molecules on the outside of the host cell.
- Some viruses have broad host ranges and can infect many different types of cells.
- Other viruses, on the other hand, have extremely narrow ranges and can infect only very specific cells.

They multiply only in living cells and it is too small to be seen individually with a light microscope. The nature, the origin, the architecture and the replication of viruses represent significant biological questions, not only for viruses themselves, but also for other living forms because learning about them contributes to better understanding of the evolution of the living world. Most plant viruses have been found in angiosperms. Relatively few viruses are known in gymnosperms, ferns, fungi or algae. Plant viruses are of great economic importance, since they cause plant diseases in a variety of crops. Virus diseases are also known in a variety of vertebrates, including fish, amphibia, birds and mammals. In invertebrates, virus diseases have been reported only in insects. Viruses have been found in practically all groups of bacteria. The host range is confined within bacterial groups, e.g. bacteriophages active in micrococci do not multiply in streptococci. Sometimes there is considerable host specificity, because of the ability of bacteria to acquire resistance by mutation. There is also evidence that cells of eukaryotic microorganisms may contain viruses. Some of these viruses may be associated with prokaryotic endosymbionts living in eukaryotic cells. Virus like particles has been observed in species of protozoa, algae and fungi. They have been observed in several protozoa lincluding *Leshmania, Entamoeba histolytica, Acanthamoeba* sp., *Naegleria, Plasmodium vivax, Plasmodium berghei Paramecium Aurelia, Carchesium polypinum and Ignotocoma sabellarum*. A virus like particles has been reported in *Plasmodium berghei.* Its structure is like that of a cytoplasmic polyhedrosis virus, a dsRNA virus of insects. There is evidence for two ds DNA viruses in *Entamoeba histolytica*. These particles have been also reported in *Aulacomonas submarina, Chara corralina, Oedogonium sp., Platymonas sp., Radiofilum sp., Uronema gigas*, and a few other algae. Bacteriophages like virus particles have been found in *Chlorella* and have been called chlorellophages. The tobacco mosaic virus has been reported in diseased plants of *Chara corralina*. The particles contain ssRNA. dsDNA viruses have been detected in cells of *Penicillium* moulds. dsRNA viruses are widely present in the higher fungi. They are predominantly latent, relatively small, and spherical to polyhedral particles with simple capsid. The name mycornaviruses has been suggested for fungal viruses containing dsRNA. The bacilliform particle of *Agaricus bisporous* is the only fungal virus reported that contains ssRNA. Viruses do not divide and do not produce any kind of specialized reproductive structures. Instead, they multiply by inducing host cells to form more viruses. Viruses do not cause disease by consuming cells or killing them with toxins, but by utilizing cellular substances during multiplication, taking up space in cells, and disrupting cellular processes.

Nature of Virus

The nature of viruses wasn't understood until the twentieth century, but their effects had been observed for centuries. British physician Edward Jenner even discovered the principle of inoculation in the late eighteenth century, after he observed that people who contracted the mild cowpox disease were generally immune to the deadlier smallpox disease. By the late nineteenth century, scientists knew that some agent was causing a disease of tobacco plants, but would not grow on an artificial medium (like bacteria) and was too small to be seen through a light microscope. Advances in live cell culture and microscopy in the twentieth century eventually allowed scientists to identify viruses. Advances in genetics dramatically improved the identification process. All viruses are obligate intracellular parasites. They take over a host cells synthetic machinery in order to reproduce and be transmitted to other cells. The basic life-process of viruses, multiplication, which means infectivity and parasitism at the same time, and consequently leads to chaos in infected cells, is also brought to the perfection. How otherwise to explain the process which happens at level of the most responsible cellular functions, by using cellular energy, cellular components, the mechanisms of the synthesis of proteins and nucleic acids and, on the other hand, leads to the collapse of a cell itself. And how to explain that viruses, with limited genetic information and without their own metabolism, can direct such perfect machinery as a cell to the synthesis of its own parts.

General Lifecycle of Virus

- The general life cycle of a virus can be described by:
  1. Recognition of host
  2. Genetic material enters host
  3. Replication using host nucleotides
  4. Protein synthesis using host enzymes, ribosomes, trna, atp
  5. Self-assembly of capsids and packaging of genome
  6. Release from host
- There are many variations of this cycle depending on the type of virus and the host.

This process has been brought to the perfection probably in order to enable the existence of viruses as a species. Such a perfect process as the multiplication of viruses makes chaos in an infected cell, which is manifested in the derangement of its structure and function, and which results in various cellular deformities and disintegration. Viruses have brought all the processes necessary for their own existence into such harmony that making chaos in the cells of attacked organisms, they do not annihilate and destroy themselves. By this chaos viruses produce infectious progeny and prolong their biological function. As metabolically inert and energetically dependant, viruses have found the unique way of their replication. Although many aspects of their replication in vivo (for example - the sites, time and mechanisms by which infecting viral genomes become uncoated, the mechanisms by which an infecting nucleic acid organizes characteristic replication sites within the cell, the mechanisms of modification of cell's organelles, etc.) have already been brought to light so far, there is still the question of how viruses initiate a cell to form as much as possible new particles by itself, coming to its own chaos during that.

2

STRUCTURE OF VIRUS

Viruses are very small, infectious, obligate intracellular molecular parasites, which do not respire, move or grow. The virus genome is composed either of DNA or RNA and directs the viral replication by the synthesis of virion components within an appropriate host cell. De novo assembly from newly synthesized components within the host cell forms progeny viruses. The size of viruses is variable. Most viruses are much smaller than bacteria. The larger viruses are, however, as large as some of the smaller bacteria. Viruses are measured in terms of nanometers (nm). Their size range is varying between 20-300 nm. They are visible only with the aid of an electron microscope. The unique properties of a virus are contained in the definition given by Agrios. He defines a virus is a nucleoprotein that has the ability to cause disease. It is noticeable that it places no requirements on the nucleoprotein entity for reproduction or synthesis. In most cases these qualities are coded for, or directed by the nucleoprotein, but they are not part of the definition of a virus. Also noticeable that no greater simplicity or complexity is required, though often viruses are more complex than "simple" nucleoproteins. Neither the type of nucleic acid (RNA or DNA), strandedness (double or single), type or number of proteins in the nucleocapsid, nor the type of nucleocapsid (rod or sphere) whether it is encompassed by an envelope or has appendages are included in the definition; though they are definitely important for each virus. Viriods, Virusoids and Prions are some virus-like particles. Viroids are small (200-400nt), circular RNA molecules with a rod-like secondary structure, which possess no capsid or envelope, which are associated with certain plant diseases. Their replication strategy like that of viruses - they are obligate intracellular parasites. Virusoids are satellite, viroid-like molecules, somewhat larger than viroids (e.g. approximately 1000nt), which are dependent on the presence of virus replication for multiplication (hence 'satellite'), they are packaged into virus capsids as passengers. Prions are rather ill defined infectious agents believed to consist of a single type of protein molecule with no nucleic acid component. Confusion arises from the fact that the prion protein and the gene, which encodes it, are also found in normal 'uninfected' cells. These agents are associated with diseases such as Creutzfeldt-Jakob disease in humans, scrapie in sheep and bovine spongiform encephalopathy (BSE) in cattle.

History about Virus Structure

The beginning of the history of structural virology is not so clearly defined, but it can be arbitrarily chosen to start in 1935 with Stanley's crystallization of TMV. This original Nobel Prize-winning work was followed by crystallization of other plant viruses including the crystallization of tomato bushy stunt virus by Bawden and Pirie. It seems rather obvious that the plant viruses would lead the way in these studies because it was possible to obtain large quantities of purified virus - an essential starting point for obtaining crystals for X-ray diffraction. Obtaining crystals is just a first small step in the goal of determining a structure. In 1956, Francis Crick and James Watson proposed that: "a small virus contains identical sub-units, packed together in a regular manner". The idea that a small virus is composed of a shell of protein that consists of a "large number of identical small protein molecules" and that "these small protein molecules aggregate around the ribonucleic acid in a regular manner. Based on possible packing arrangements, they also predicted that "for a spherical virus the number of sub-units is likely to be a multiple of 12". Another virologist, Don Caspar published two precession photographs obtained from X-ray studies of tomato bushy stunt virus and wrote "the X-ray evidence shows that bushy stunt virus possesses subunits. The number of sub-units is certainly a multiple of 12 and very probably a multiple of sixty." In1962, Caspar and Klug were published some of the fundamental ideas of the structure of icosahedral viruses. They discuss the concept of quasi-equivalence and acknowledge that the principles set down by Buckminster

Fuller for the construction of geodesic domes were an inspiration for the development of their ideas. The interest in and importance of this work may not have had many ramifications for virologists - diffraction patterns don't help explain how a virus can enter a cell or assemble into an infectious particle. Some of the questions being resolved by virologists at that same meeting included recombination in RNA viruses. The paper presented by George Hirst was "Genetic recombination with Newcastle Disease Virus, Polioviruses and Influenza". Recombination was not observed with Newcastle Disease virus. The structure of the RNA genome of influenza virus was not yet clear, but the high frequency of recombination led Hirst to suggest "the RNA from this agent is often in fragments". The paper by Franklin and Baltimore ("Patterns of macromolecular synthesis in normal and virus-infected cells") discussed the possibility that a new RNA dependent RNA polymerase was induced in cells infected by Mengovirus and that the enzyme was coded by the viral genome. A significant number of the talks were concerned with transformation of cells by viruses and the characterization of those cells. In his concluding remarks Dulbecco stated a major problem of that period: "In the transformation caused by DNA-containing viruses the most outstanding event is the ultimate disappearance of the virus itself. It is most likely that the virus causing the transformation is never recoverable in infectious form from the cell it infected. This is an unfortunate event, because it renders extremely difficult if not impossible the isolation of viral mutants causing special types of transformation".

General Structure of Virus

Technically, Viruses are not really organisms. They are very small in size and have a simple, but effective structural organization. Viruses usually consist of just two or three categories of components and use the components of the Host Cell to perform their "metabolism". This makes them a special type of "parasite". Generally, viral structure has three components:

Capsid — Protein Coat

Genome — DNA or RNA

Envelope — (not all viruses)

Capsid

The capsid is the protein shell that encloses the nucleic acid; with its enclosed nucleic acid, it is called the nucleocapsid. This shell is composed of protein organized in subunits known as capsomers. They are closely associated with the nucleic acid and reflect its configuration, either a rod-shaped helix or a polygon-shaped sphere. The capsid has three functions: 1) it protects the nucleic acid from digestion by enzymes, 2) contains special sites on its surface that allow the virion to attach to a host cell, and 3) provides proteins that enable the virion to penetrate the host cell membrane and, in some cases, to inject the infectious nucleic acid into the cell's cytoplasm. Under the right conditions, viral RNA in a liquid suspension of protein molecules will self-assemble a capsid to become a functional and infectious virus.

Envelope

Many types of virus have a glycoprotein envelope surrounding the nucleocapsid. The envelope is composed of two lipid layers interspersed with protein molecules (lipoprotein bilayer) and may contain material from the membrane of a host cell as well as that of viral origin. The virus obtains the lipid molecules from the cell membrane during the viral budding process. However, the virus replaces the proteins in the cell membrane with its own proteins, creating a hybrid structure of cell-derived lipids and virus-derived proteins. Many viruses also develop spikes made of glycoprotein on their envelopes that help them to attach to specific cell surfaces.

Some Terminology

Virion : Intact virus or infectious unit. Free virus particle.

Capsid : Protein shell arranged around the N.A., or complexed to the nucleic acid. Capsid is often called the protein coat. These are usually repeat units, and damage to one such unit does not render the virus non-viable.

Nucleocapsid : Complete protein nucleic acid complex that is the packaged form of the genome inside the envelope, or may be naked, that is non-enveloped.

Capsomere : Morphological subunits that consist of identical or different proteins molecules. The capsid is made up of capsomeres. Another term used is protein subunits. These are held together by non covalent bonds.

Envelope : Lipid rich outer layer of virus, derived from host cell. also referred to as peplos.

Peplomers: Spikes from outer envelope.

Genome or nucleic acid

Just as in cells, the nucleic acid of each virus encodes the genetic information for the synthesis of all proteins. While the double-stranded DNA is responsible for this in prokaryotic and eukaryotic cells, only a few groups of viruses use DNA. Most viruses maintain all their genetic information with the single-stranded RNA. There are two types of RNA-based viruses. In most, the genomic RNA is termed a plus strand because it acts as messenger RNA for direct synthesis (translation) of viral protein. A few, however, have negative strands of RNA. In these cases, the virion has an enzyme, called RNA-dependent RNA polymerase (transcriptase), which must first catalyze the production of complementary messenger RNA from the virion genomic RNA before viral protein synthesis can occur.

Shape of Virus

There are predominantly two kinds of shapes found amongst viruses: rods, or filaments, and spheres. The rod shape is due to the linear array of the nucleic acid and the protein subunits making up the capsid. The sphere shape is actually a 20-sided polygon (icosahedron). Although they may appear pleomorphic, this is due to the lipid bilayer membrane envelope found around most viruses. In 1956 Watson and Crick, tried to clear why do viruses have two basic shapes? They proposed that the capsid of small viruses would have to be built from identical subunits. This hypothesis was based on the low molecular weight of viral nucleic acid. A viral genome is not large enough to code for many genes. e.g. RNA of polio has a molecular wt. of 2×10^6. A triplet codon has a mol wt. of app. 1000 $=1 \times 10^3$ therefore enough RNA for 2×10^3 amino acids. The average protein has 300-500 amino acids. Therefore dividing 2×10^3 by $3 \times 10^2 = 6.6$ proteins. If it were a DNA virus would have to take double stranded ness into account and thus enough DNA for 3 proteins. This assumes that all the nucleic acids usable. Thus viral capsids often consist of repeating subunits of the same protein. This implies a very efficient structure, subunit proteins have a tendency to bind to each other, and thus the information for morphogenesis lies in the amino acid structure of the virus. Thus most virus structures undergo self-assembly. The assembly of viruses from small repeating subunits has several advantages to the virus: it reduces the amount of nucleic acid required, and it minimizes the risk of incurring fatal errors, as faulty subunits can be discarded. However, the proteins are non-symmetrical and yet the virus forms regular geometric shapes. The high degree of quaternary structure in virus capsids explains their stability and resistance to various agents and enzymes. The multiple bonds formed during the aggregation of protein subunits contributes to the masking of enzyme-susceptible sites, and to stabilization against heat and other agents. In 1955 Fraenkel Conrat and Williams showed self-assembly of mixtures of TMV coat protein and TMV RNA. Thus virus particles could form spontaneously from purified subunits without any extraneous material. The forces, which drive self-assembly, are hydrophobic and electrostatic interactions, only rarely are co-valent bonds involved in virus assembly. In biological terms this is protein-protein, protein-nucleic acid and protein-lipid interactions.

Capsid

The capsid denotes the protein shell that encloses the nucleic acid. It is built of structure units. Structure units are the smallest functional equivalent building units of the capsid. Capsomers are morphological units seen on the surface of particles and represent clusters of structure units. The capsid together with its enclosed nucleic acid is called the nucleocapsid. The nucleocapsid may be invested in an envelope, which may contain material of host cell as well as viral origin. It composed of lipids, proteins and carbohydrates. The envelope may also have spikes projecting from the surface. Whether a virus has an envelope or spikes is again determined by the nucleic acid. Capsid is a complex and highly organized entity, which gives form of the virus. The complex arrangements of macromolecules in the capsid are minute marvels of molecular architecture. Specific requirements of each type of virus have resulted in a fascinating apparent diversity of organization and geometrical design. Nevertheless, there are certain common features and general principles of architecture that apply to all viruses.

In 1956, Crick and Watson proposed on theoretical considerations and on the basis of rather flimsy experimental evidence then available, principles of virus structure that have been amply confirmed and universally accepted. They first pointed out that the nucleic acid in small virions was probably insufficient to code for more than a few sorts of protein molecules of limited size. The only reasonable way to build a protein shell, therefore, was to use the same type of molecule over and over again, hence their theory of identical subunits.The second part of their proposal concerned the way in which the subunits must be packed in the protein shell or capsid. On general grounds it was expected that subunits would be packed so as to provide each with an identical environment. This is possible only if they are packed symmetrically. Crick and Watson pointed out that the only way to provide each subunit with an identical environment was by packing them to fit some form of cubic symmetry. A body with cubic symmetry possesses a number of axes about which it may be rotated to give a number of identical appearances. These predictions were soon confirmed and it became evident that the occurrence of icosahedral features in quite unrelated viruses was not a matter of chance selection but that icosahedral symmetry is preferred in virus structure. Different viruses have different proteins on their surfaces, which interact with specific receptors on their host cells. The specificity of the reaction between viral protein and host receptor defines and limits the host species as well as the type of cell that is infected. The function of the capsid of a virus particle is to protect the fragile nucleic acid genome from following damages.

- Physical damage - Shearing by mechanical forces.
- Chemical damage - UV irradiation (from sunlight) leading to chemical modification.
- Enzyme damage - Nucleases derived from dead or leaky cells or deliberately secreted by vertebrates as defence against infection.

Protein subunits in a virus capsid are multiply redundant, i.e. present in many copies per particle. Damage to one or more subunits may render that particular subunit non-functional, but does not destroy the infectivity of the whole particle.

The orientation and location of the 240 hexons comprising the outer protein shell of adenovirus have been determined by Burnett (1985). Electron micrographs of the capsid and its fragments were inspected for the features of hexon known from the X-ray crystallographic model as described in his paper. A capsid model is proposed with each facet comprising a small p3 net of 12 hexons, arranged as a triangular sextet with three outer hexon pairs. The sextet is centrally placed about the icosahedral threefold axis, with its edges parallel to those of the facet. The outer pairs project over the facet edges on one side of the icosahedral twofold axes at each edge. The model capsid is defined by the underlying icosahedron, of edge 445 Å, upon which hexons are arranged. The hexons are thus bounded by icosahedra with insphere radii of 336 Å and 452 Å. A quartet of hexons forms the asymmetric unit of an icosahedral hexon shell, which can be closed by the addition of pentons at the 12 vertices. Considering the hexon trimer as a complex structure unit, its interactions in the four topologically distinct environments are very similar, with conservation of at least two-thirds of the inter-hexon bonding. The crystal-like construction explains the flat facets and sharp edges characteristic of adenovirus. Larger "adenovirus-like" capsids of any size could be formed using only one additional topologically different environment. The construction of adenovirus illustrates how an impenetrable protein shell can be formed, with highly conserved intermolecular bonding, by using the geometry of an oligomeric structure unit and symmetry additional to that of the icosahedral point group. This contrasts with the manner suggested by Caspar and Klug (1962), in which the polypeptide is the structure unit, and for which the number of possible bonding configurations required of a structure unit tends to infinity as the continuously curved capsid increases in size. The known structures of polyoma and the plant viruses with triangulation number equal to 3 are evaluated in terms of hexamer-pentamer packing, and evidence is presented for the existence of larger subunits than the polypeptide in both cases. It is suggested that spontaneous assembly can occur only when exact icosahedral symmetry relates structure units or sub-assemblies, which would themselves have been formed by self-limiting closed interactions. Without such symmetry, the presence of scaffolding proteins or nucleic acid is necessary to limit aggregation.

The proper assembly of proteins and nucleic acids into biologically active virions involves numerous and diverse macromolecular interactions. While structural proteins must correctly interact, proper morphogenesis is equally dependent on scaffolding proteins, which are transiently associated with nascent protein complexes during virion assembly. Scaffolding proteins can also mediate conformational switches: structural changes that cause the next assembly process to differ from the previous one. The atomic structures of the øX174 virion and procapsid, containing two scaffolding proteins (internal and external scaffolding proteins), have been solved by McKenna et al., 1992, 1994; Dokland et al., 1997, 1999. Scaffolding proteins are transiently associated with morphogenetic intermediates but not found in the mature particle (King and Casjens, 1974). While the COOH-termini of the øX174 internal scaffolding protein is readily distinguished within the crystal structure, much of the amino-terminus is unordered; suggesting that interactions with the overlying coat protein can be both variable and flexible. To prove this, the internal scaffolding genes of øX174, G4 and alpha3 were cloned and expressed in vivo. Although the amino acid sequences of these proteins exhibit 70% divergence in their primary structures, they can productively direct the assembly of viral coat proteins in a non-species specific manner (Burch et al., 1999). However, cross-complementation is not as efficient as complementation by the indigenous protein. Similar experiments conducted with chimeric proteins indicate that complementation efficiency appears to be governed by the COOH-termini of the proteins (Burch and Fane, 2000). Similar results have been obtained with experiments conducted on both the P22 and Herpes scaffolding proteins (Parker et al., 1998; Preston et al., 1997).

It has also investigated that the functions of the amino-terminus of the internal scaffolding protein, which is unordered in within the atomic structure, by generating nested sets of deletions in the gene. While proteins missing the first 60 amino acids appear to have no activity in complementation assays, proteins lacking as much as 42 amino acids can complement some amber mutants. However, the ability to complement may depend on the length of the amber fragment used in the experiment, suggesting a subunit complementation phenomenon. The internal scaffolding protein may also contain auto-proteolytic activity. The atomic structure of the external scaffolding protein, protein D, suggests that it also shares many features with molecular chaperones (Dokland et al., 1997, 1999). The four D proteins per asymmetric unit have different structures (D1, D2, D3, and D4) and are arranged as two similar, but not identical, dimers (D1D2 and D3D4). The D1D2 dimer makes contact with the major spike protein, protein G. The D3D4 dimer makes contact with the major coat protein, protein F. Like a molecular chaperone (Hartl and Martin, 1995) protein D can bind to other proteins in many different ways. Its structure elucidates the conformational requirements needed for the pliable and diverse interactions of chaperone-like molecules with their substrates. Capsid is usually a polyhedron (usually icosahedral), spiral (helical symmetry), or it may be more complex.

Genome

The composition and structure of virus genomes (i.e. the nucleic acid which encodes the genetic information of the virus) is more varied than any of those seen in the entire bacterial, plant or animal kingdoms. The nucleic acid comprising the genome may be single-stranded or double-stranded and in a linear, circular or segmented configuration. Virus genomes range in size from approximately 3,500 nucleotides (nt) (e.g. bacteriophages of the family Leviviridae such as MS2 and Qb) to approximately 230 kilobase pairs (kbp). Unlike the genomes of all cells, which are composed of DNA, virus genomes may contain their genetic information encoded in either DNA or RNA. Whatever the particular composition of a virus genome, all must conform to one condition. Since viruses are obligate intracellular parasites only able to replicate inside the appropriate host cells, the genome must contain information encoded in a form, which can be recognized and decoded by the particular type of cell parasitized. Thus, the genetic code employed by the virus must match or at least be recognized by the host organism. Similarly, the control signals, which direct the expression of virus genes, must be appropriate to the host.

Many of the DNA viruses of eukaryotes closely resemble their host cells in terms of the biology of their genomes:

- Some DNA virus genomes are complexed with cellular histones to form a chromatin-like structure inside the virus particle.
- Vaccinia virus mRNAs were found to be polyadenylated at their 3' ends by Kates in 1970 - the first observation of this phenomenon.
- Split genes containing non-coding introns, protein coding exons and spliced mRNAs were first discovered in adenoviruses by Sharp in 1977.

As already described, the new techniques of molecular biology have had a major influence in concentrating much attention on the virus genome. Initially, the questions to be asked about any virus genome will usually include the following:

- Composition - DNA or RNA, single-stranded (ss) or double-stranded (ds), linear or circular.

- Size and number of segments.
- Terminal structures.
- Nucleotide sequence.
- Coding capacity - open reading frames.
- Regulatory signals - transcription enhancers, promoters and terminators.

Direct analysis by electron microscopy, if calibrated with known standards, can be used to estimate the size of nucleic acid molecules.The most important single technique has been gel electrophoresis. It is most common to use agarose gels to separate large nucleic acid molecules (several megabases or kilobases) and polyacrylamide gel electrophoresis (PAGE) to separate smaller pieces (a few hundred bp down to a few nucleotides).

Nucleotide sequencing is dependent on the ability to separate molecules, which differ from each other by only one nucleotide in size. The relative simplicity of virus genomes (compared with even the simplest cell) offers a major advantage - the ability to 'rescue' infectious virus from purified or cloned nucleic acids. Infection of cells caused by nucleic acid alone is referred to as transfection:

Single-stranded virus genomes may be:

- positive (+) sense, i.e. of the same polarity (nucleotide sequence) as mRNA
- negative (-) sense
- ambisense - a mixture of the two.

Virus genomes which consist of (+) sense RNA (i.e. the same polarity as mRNA) are infectious when the purified vRNA is applied to cells in the absence of any virus proteins. This is because (+) sense vRNA is essentially mRNA and the first event in a normally infected cell is to translate the vRNA to make the virus proteins responsible for genome replication. In this case, direct introduction of RNA into cells merely circumvents the earliest stages of the replicative cycle.

In most cases, virus genomes, which are composed of double-stranded DNA, are also infectious. The events, which occur here, are a little more complex, since the virus genome must first be transcribed by host polymerases to produce mRNA. Using these techniques, virus can be rescued from cloned genomes, including those, which have been manipulated in vitro.

The following generalizations may be made about viral genome.

- Bacterial viruses contain ssDNA, dsDNA, ssRNA or dsRNA. Most bacteriophages contain only DNA. MS2 contains ssRNA and 6 contains dsRNA.
- Animal viruses contain dsDNA, ssRNA or dsRNA. They do not contain ssDNA.
- Plant viruses usually contain only RNA (ssRNA or dsRNA).

RNA Virus Genomes

Single stranded RNA (ssRNA) virus genome

(a) *Positive-Strand ssRNA Viruses*: The ultimate size of single-stranded RNA or ssRNA genomes is limited by the fragility of RNA and the tendency of long strands to break. In addition, RNA genomes tend to have higher mutation rates than those composed of DNA because they are copied less accurately, which also tends to drive RNA viruses towards smaller genomes. Single-stranded RNA genomes vary in size from those of Coronaviruses at approximately 30kb long to those of bacteriophages such as MS2 and Qb at about 3.5kb. Such genomes from different virus families share a number of common features:

- Purified (+) sense vRNA is directly infectious when applied to susceptible host cells in the absence of any virus proteins

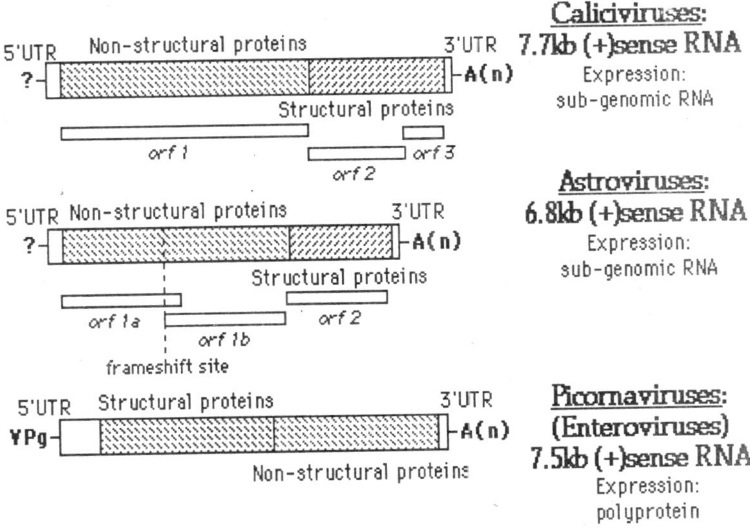

Fig. 2.1. Virus genome (positive strand ssRNA).

(although it is about one million times less infectious than virus particles).

- There is an untranslated region (UTR) at the 5' end of the genome, which does not encode any proteins, and a shorter UTR at the 3' end. These regions are functionally important in virus replication and are thus conserved in spite of the pressure to reduce genome size.
- Both ends of (+) stranded eukaryotic virus genomes are often modified, the 5' end by a small, covalently attached protein or a methylated nucleotide 'cap' structure and the 3' end by polyadenylation. These signals allow vRNA to be recognised by host cells and to function as mRNA.

(b) *Negative-Strand ssRNA Viruses*: Viruses with negative-sense RNA genomes are a little more diverse than positive-stranded viruses. Possibly because of the difficulties of expression, they tend to have larger genomes encoding more genetic information. Because of this, segmentation is a common though not universal feature of such viruses. Negative-sense RNA genomes are not infectious as purified RNA. This is because such virus particles all contain a virus-specific polymerase. The first event when the virus genome enters the cell is that the (-) sense genome is copied by the polymerase, forming either (+) sense transcripts which are used directly as mRNA, or a double-stranded molecule known either as the replicative intermediate (RI) or replicative form (RF), which serves as a template for further rounds of mRNA synthesis. Therefore, since purified negative-sense genomes cannot be directly

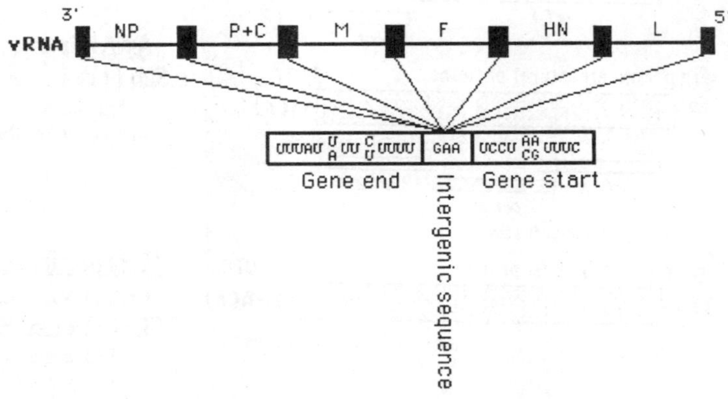

Fig. 2.2. Virus genome (negative strand ssRNA).

translated and are not replicated in the absence of the virus polymerase, these genomes are inherently non-infectious.

(c) *Ambisense Genome Organization of ssRNA Viruses*: Some RNA viruses are not strictly 'negative-sense' but ambisense, since they are part (-) sense and part (+) sense

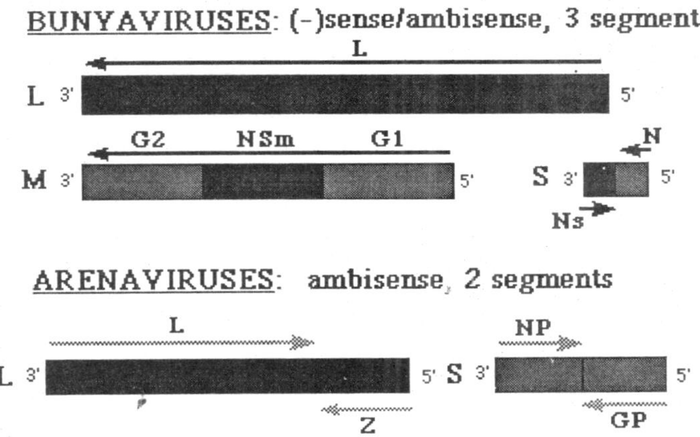

Fig. 2.3. Bunyanviruses and Arenaviruses.

Double stranded RNA (dsRNA) virus genome

Double stranded RNA is found in animal viruses like reovirus, the blue tongue virus, the cytoplasmic polyhedrosis virus of insects and retroviruses and in plant viruses like the wound tumor virus and the rice dwarf virus. In the reoviruses, the cytoplasmic polyhedrosis virus of insects and the blue tongue virus the dsRNA consists of ten or more segments. In the retrovirus there are diploid single strands forming a complex with 7S and 4S RNA of the host cell.

DNA Virus Genomes

'Small' DNA Genomes

Bacteriophages have been extensively studied as examples of DNA virus genomes. Although they vary considerably in size, in general terms they tend to be relatively small. The structure of the bacteriophage M13 genome has been studied in great detail and modified extensively for use as a vector for DNA sequencing. The genome of this virus is:

- Circular

- Single-stranded DNA
- Approximately 7,200 nucleotides long

Unlike other virion structures, the filamentous M13 capsid can be lengthened by the addition of further protein subunits. The genome size of this virus can also be increased by the addition of extra sequences in the non-essential intergenic region without the penalty of becoming incapable of being packaged into the capsid. This is very unusual. In other viruses, the packaging constraints are much more rigid, e.g. in phage lambda, only DNA of between approximately 95% - 110% (approximately 46kbp - 54kbp) of the normal genome size (49kbp) can be packaged into the virus particle.

Not all bacteriophages have such simple genomes as M13, e.g. the genome of lambda is approximately 49kbp and that of phage T4 about 160kbp double-stranded DNA. These latter two bacteriophages also illustrate another common feature of linear virus genomes - the importance of the sequences present at the ends of the genome.

In the case of lambda, the substrate, which is packaged into the phage heads during assembly, consists of long concatemers of

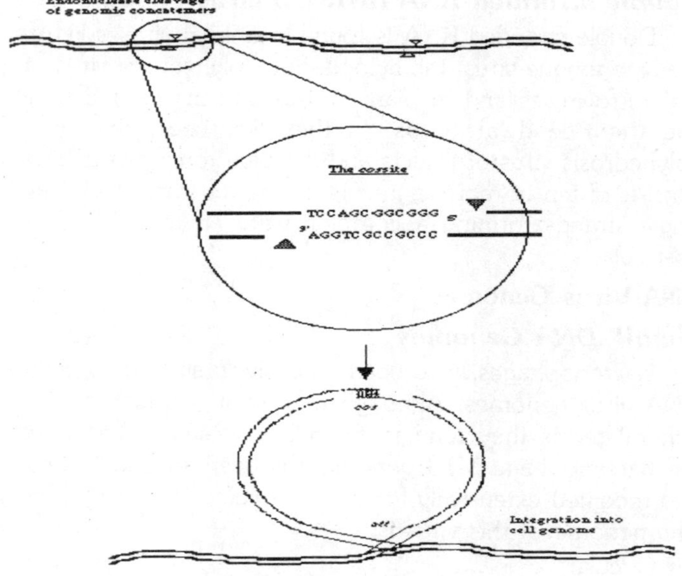

Fig. 2.4. Circular genome.

phage DNA, which are produced during the later stages of vegetative replication. The DNA is 'reeled in' by the phage head and when a complete genome has been incorporated, cleaved at a specific sequence by a phage-encoded endonuclease. This enzymes leaves a 12bp 5' overhang on the end of each of the cleaved strands, known as the cos site. Hydrogen bond formation between these 'sticky ends' can result in the formation of a circular molecule. In a newly infected cell, the gaps on either side of the cos site are closed by DNA ligase and it is this circular DNA which is undergoes vegetative replication or integration into the bacterial chromosome.

Bacteriophage T4 illustrates another molecular feature of certain linear virus genomes, terminal redundancy. Replication of the T4 genome also produces long concatemers of DNA. These are cleaved by a specific endonuclease, but unlike the lamda genome, the lengths of DNA incorporated into the particle are somewhat longer than a complete genome length. Therefore, some genes are repeated at each end of the genome and the DNA packaged into the phage particles contains reiterated information.

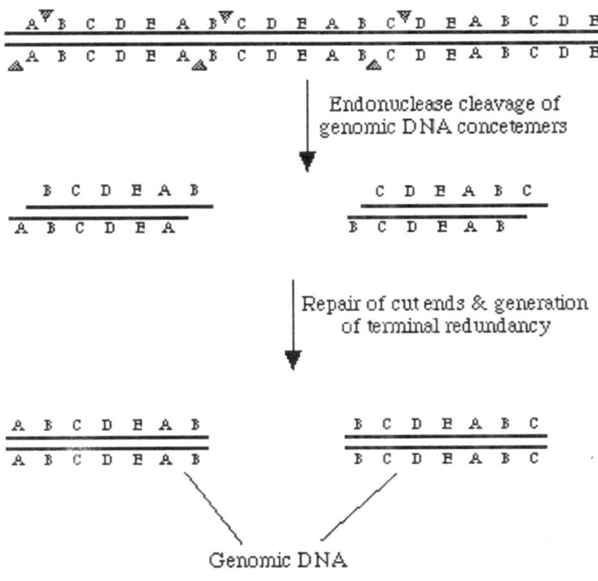

Fig. 2.5. Linear genome.

As further examples of small DNA genomes, consider those of two families of animal viruses, the parvoviruses and polyomaviruses. These are very small genomes and even the replication-competent parvoviruses contain only two genes, rep, which encodes proteins involved in transcription and cap, which encodes the coat proteins. The ends of the genome have palindromic sequences of about 115nt, which form 'hairpins'. These structures are essential for the initiation of genome replication, once again emphasising the importance of the sequences at the ends of the genome. The genomes of polyomaviruses consist of double-stranded, circular DNA molecules, approximately 5kbp in size. The entire nucleotide sequence of all the viruses in the family is known and the architecture of the polyomavirus genome (i.e. number and arrangement of genes and function of the regulatory signals and systems) has been studied in great detail at a molecular level. Within the particles, the virus DNA is associated with four cellular histones. The genomic organization of these viruses has evolved to pack maximal information (6 genes) into minimal space (5kbp). This has been achieved by the use of both strands of the genome DNA and overlapping genes.

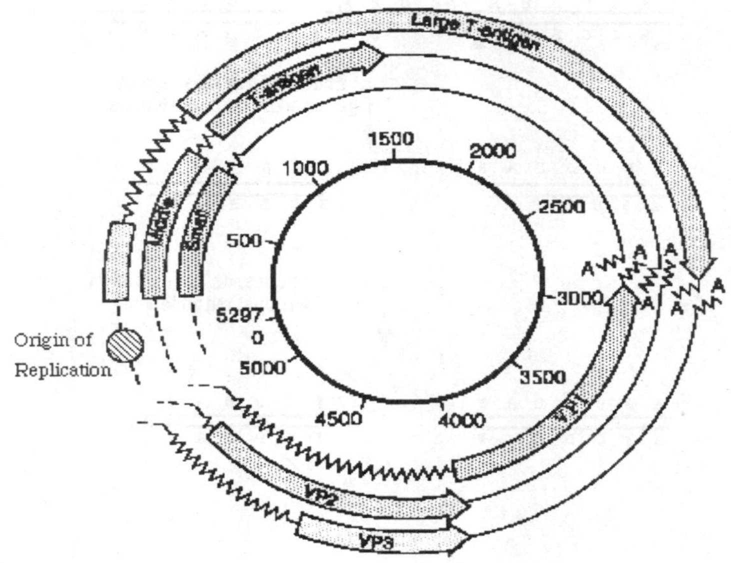

Fig. 2.6. Polymovirus genome.

'Large' DNA Genomes

There are a number of virus groups which have double-stranded DNA genomes of considerable size and complexity. In many respects, these viruses are genetically very similar to the host cells, which they infect. Two examples of such viruses are the adenovirus and herpesvirus families.

(a) *Herpesvirus genomes*: The herpesviruses are a large family containing more than 100 different members, at least one for most animal species, which have been examined to date, including seven human herpesviruses. Herpesviruses have very large genomes composed of up to 230kbp linear, double-stranded DNA. The different members of the family are widely separated in terms of genomic sequence and proteins, but all are similar in terms of structure and genome organization. Some herpesvirus genomes consist of two covalently joined sections, a unique long (UL) and a unique short (US) region, each bounded by inverted repeats. The repeats allow structural rearrangements of the unique regions and therefore, these genomes exist as a mixture of four isomers, all of which are functionally equivalent. Herpesvirus genomes also contain multiple repeated sequences and depending on the number of these, the genome size of various isolates of a particular virus can vary by up to 10kbp.

(b) *Adenovirus genomes*: The genomes of adenoviruses consist of linear, double-stranded DNA of 30-38kbp. These viruses contain 30-40 genes. The terminal sequences of each DNA strand are inverted repeat of 100-140bp and therefore, the denatured single strands can form 'panhandle' structures. These structures are important in DNA replication. Although adenovirus genomes are considerably smaller than those of herpesviruses, the expression of the genetic information is rather more complex. Clusters of genes are expressed from a limited number of shared promoters. Multiply spliced mRNAs and alternative splicing patterns are used to express a variety of polypeptides from each promoter. In contrast, herpesvirus genes each tend to be expressed from their own promoter - resulting in a much larger genome.

Segmented and Multipartite Virus Genomes

Segmented virus genomes are those, which are divided into two or more physically separate molecules of nucleic acid, all of which are then packaged into a single virus particle. Multipartite genomes are those which are segmented and where each genome segment is packaged into a separate virus particle. These discrete particles are structurally similar and may contain the same component proteins, but often differ in size depending on the length of the genome segment packaged.

8 segments negative-sense RNA

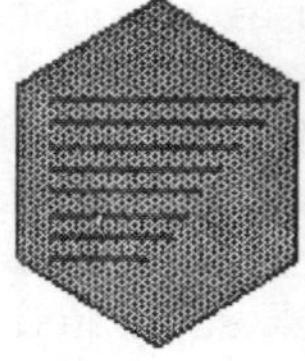

Segmented

Fig. 2.7. Influenza virus.

2 molecules double-stranded DNA

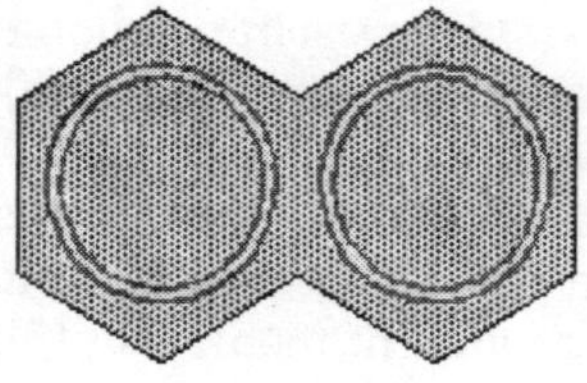

Bipartite

Fig. 2.8. Gemini virus.

There are many examples of segmented virus genomes, including many human, animal and plant pathogens such as orthomyxoviruses, reoviruses and bunyaviruses. There are rather fewer examples of multipartite viruses, all of which infect plants. These include:

(a) Bipartite viruses: They have two genome segments/virus particles.

(b) Tripartite viruses: They have three genome segments/virus particles.

Separating the genome segments into different particles removes the requirement for accurate sorting, but introduces a new problem in that all of the discrete virus particles must be taken up by a single host cell to establish a productive infection. This is perhaps the reason why multipartite viruses are only found in plants. Many of the sources of infection by plant viruses, such as inoculation by sap-sucking insects or after physical damage to tissues, results in a large input of infectious virus particles, providing the opportunity for infection of an initial cell by more than one particle.

Envelope

Envelope is an amorphous structure composed of lipid, protein and carbohydrate, which lie to the outside of the capsid. It contains a mosaic of antigens from the host and the virus. A naked virus is one without an envelope. Enveloped viruses have glyco-protein spikes arising from their envelope. These spikes have enzymatic, absorptive, hemagglutinating and/or antigenic activity. The envelope contains host and viral proteins that are important for interaction with cellular components during the process of infection and replication. The host's defense mechanisms (cellular and humoral mediated responses) are directed against these viral antigenic epitopes. Envelope is also reffered to as peplos. Most viruses are enveloped. Only seven families of virus are non-enveloped. The envelope is derived from the plasma membrane, endoplasmic reticulam, golgi body or nuclear membrane. Contains a lipid bilayer and proteins with special functions, usually glycoproteins. There are two kinds of viral proteins, found as part of the envelope, one or more glycoproteins and a matrix protein. In arboviruses, rhabdoviruses and myxoviruses, there is overwhelming evidence that all envelop proteins are coded by viral genomes. In leukoviruses, herpesviruses and poxyviruses, it is not clear whether host cell proteins are present in the envelop or not. These viruses, however, do contain proteins specified exclusively by virus. The envelops of rhabdoviruses, orthomyxoviruses and paramyxoviruses contain one major carbohydrate-free polypeptide. This is the smallest polypeptide of all the structural proteins of the virion (molecular Weight 20,000-30,000 in rhabdo.viruses and orthomyxoviruses and about 40,000 in paramyxoviruses). In the influenza virus the carbohydrate-free protein is the main protein, comprising 50% of the envelope protein and 34-40% of the entire virion protein. Rhabdoviruses contain additional carbohydrate-free proteins. The membranes of all classes of enveloped viruses contain glycoproteins. The group A arboviruses contain a single protein in their envelops. This protein is a glycoprotein. In the Sindbis virus and the Semliki Forest virus the protein contain a relatively high proportion of hydrophobic amino acids, indicating that it is associated with the envelope lipids. Rhabdoviruses have one glycoprotein in their envelopes, paramixoviruses two glycoproteins and influenza viruses (orthomyxoviruses) four different glycoproteins. The herpesviruses and the leukoviruses also have glycoproteins in their envelopes. The spikes on the outer surface of virions are glycoproteins. The glycoproteins are anchored in the lipid bilayer by means of hydrophobic bonds; only about 30 amino acids penetrate into the virus. Thus these proteins have a large external domain and a small cytoplasmic domain. They are transmembrane proteins. The viral envelopes are modified host membranes. These glycoproteins are oligomers that are associated with each other to form tetramers etc. There are often the spikes seen on the virus surface. Some glycoproteins such as the influenza hemaglutinin are anchored at both ends, thus forming a loop. In such cases a signal sequence at the amino end has been removed. There are two types of glycoproteins. External glycoprotein anchored in the envelope by a single transmembrane domain, and a short internal tail. These proteins are usually the major antigens of the virus and involved in functions such as hemmagglutination, receptor binding, and membrane fusion. The other class is channel proteins, which are mostly hydrophobic proteins that form a protein-lined channel through the envelope. This protein alters permeability of the membrane (e.g. ion channel). Such proteins are important in modifying the internal environment of the virus. A second class of proteins is matrix proteins. These are virus-encoded proteins that are found inside the cell membrane and may transverse the cell membrane. These proteins link the internal nucleocapsid to the envelope. In retroviruses this structure constitutes 30% weight of the virus. Unusual structure found in HSV is the tegument, it is in between the two membranes, and contains viral protein. glycoprotein and matrix protein. Viral envelopes contain a significant amount of carbohydrates. Galactose, mannose, glucose, fucose, glucosamine and galactosamine have been found in the influenza virus, the parainfluenza virus SV5 and in the Sindbis virus. The total carbohydrate content and the proportions of hexoses and hexosamines are very similar in these viruses.

Morphology or Symmetry in Virus

The introduction of negative staining by Brenner and Horne (1959) revolutionized the field of electron microscopy of viruses. Within just a few years, much new and exciting information about the architecture of virus particles was acquired. Not only were the overall shapes of particles revealed but also the symmetrical arrangement of their components. This led to a need for a new terminology to describe the viral components. Lwoff, Anderson and Jacob (1959) proposed the terms "capsid" and "capsomers" to represent, respectively, the protein shell and the units comprising it, and the term"virion" to denote the complete infective virus particle. This terminology was generally accepted although it later proved to be inadequate. The morphology of a virus is determined by the arrangement of the protomeres (protein subunits make capsid). When protomeres aggregate into units of five or six (capsomeres) and then condense to form a geometric figure having 20 equal triangular faces and 12 apices, the virus is said to have icosahedral (polyhedral) morphology. When protomeres aggregate to form a capped tube, they are said to have helical (cylindrical) morphology. Any other arrangement of the protomeres results in a complex morphology. So, viruses occur in three main symmetries, Icosahedral ymmetry, Helical Symmetry and Complex Symmetry.

Icosahedral Symmetry

Viruses not truly spheres, but polyhedrons. A polyhedron is a solid, bounded by planes polygons. The alternative way to build virus capsid is to arrange protein subunits in the form of a hollow sphere, enclosing the genome within. Icosahedral symmetry is identical to cubic symmetry. In electron micrographs, capsomeres are recognized as regularly spaced rings with a central hole. An icosahedron is composed of 20 facets, each an equilateral triangle, and 12 vertices, and because of the axes of rotational symmetry is said to have5:3:2 symmetry Axes of Symmetry. There are, in fact, six 5-fold axes of symmetry passing through the vertices, ten 3-fold axes extending through each face and fifteen 2-fold axes passing through the edges of an icosahedron.

Icosahedral symmetry requires definite numbers of structure units to complete a shell. According to Crick and Watson (1956), asymmetrical protein subunits packed in such a way that each has an identical environment, pointed out that a virus with 5:3:2 symmetry required a multiple of 60 subunits to cover the surface completely. Each unit would be related identically and asymmetrically with its neighbours, and none of the units would coincide with an axis of symmetry.

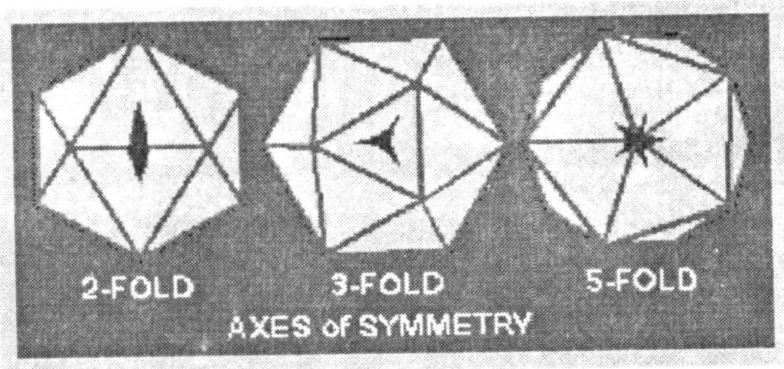

Fig. 2.9. Icosahedral symmetry or Icosahedral (isometric) symmetry

A number of solid shapes can be constructed from repeated subunits:

- Tetrahedron (4 triangular faces)
- Cube (6 square faces)
- Octahedron (8 triangular faces)
- Dodecahedron (12 pentagonal faces)
- Icosahedron (20 triangular faces)

The first high-resolution micrographs of negatively stained icosahedral viruses were obtained (Horne et al., 1959 - adenovirus; and Huxley and Zubay, 1960 - turnip yellow mosaic virus) it seemed that there was a structural paradox. The number of morphological units observed on the surface of known icosahedral viruses at that time was never 60 or multiples of 60, and was often more than 60. Furthermore, the capsomers themselves appeared to be symmetrical and were located on symmetry axes, e.g. herpesvirus. There was direct evidence that capsomers of herpesvirus were hexagonal and pentagonal in section. It is obvious that five-fold capsomers must be located on axes of five-fold symmetry, and six-fold capsomers may be situated on axes of two-fold or three-fold symmetry, or in indifferent sites where they are suited to hexagonal packing. It was therefore clear that the capsomers were not equivalent to the subunits of Crick and Watson (1956).

Supposing that the symmetrical capsomers are built from a number of asymmetrical subunits provided an obvious solution to the problem. In this way it is possible to build a variety of complicated bodies in which 5:3:2 symmetry is preserved and in which the number of subunits is a multiple of 60 as predicted by Crick and Watson.

The theoretical basis for the structure of isometric viruses was put on a firm foundation by Caspar and Klug (1962) with their concept of identical elements in quasi-equivalent environments. They defined all possible polyhedra in terms of structure units. The icosahedron itself has 20 equilateral triangular facets and therefore 20T structure units where T is the triangulation number given by the rule:

T=Pf where P can be any number of the series 1,3,7,13,19,21,31..(=h + hK +K , for all pairs of integers, h and K having no common factor) and f is any integer.

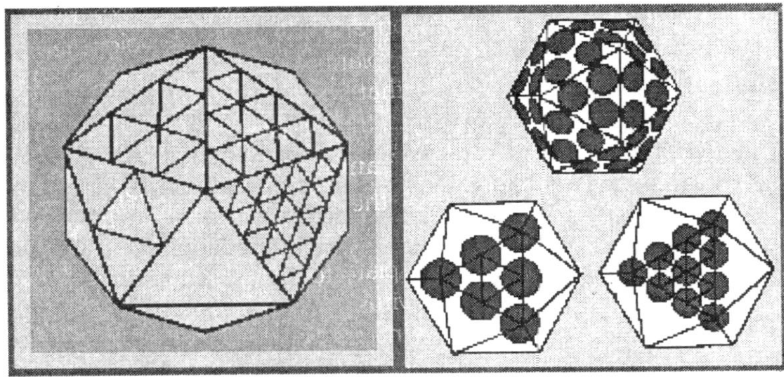

Fig. 2.10. Icosahedral symmetry.

Morphological units can be clustered as 20T trimers, 30T dimers or separated as 60T monomers. The number of morphological units that would be produced by a clustering into hexamers and pentamers can be calculated as follows: There are 10(T-1) hexamers plus 12, (And only 12) pentamers. Caspar and Klug (1962) claimed that most icosahedral viruses fall into2 classes:- P=1 and P=3; and that all deltahedra for which P=>7 are skew, and therefore exist in right and left- handed forms. One "hand" might be selected by the nucleic acid, but there would still be the chance that mistakes in assembly leading to defective particles might occur. The most probable mistake in assembly would be the formation of tubular forms. Tubular structures, which have a diameter, and surface structure similar to icosahedral virus particles have been observed associated with polyoma and papilloma viruses. An unexpected discovery has resulted from the determination during the last decade from high-resolution three-dimensional structures for a number of vertebrate, plant and insect viruses. One of the interesting facts to emerge from X-ray crystallography is the similarity in structure between the spherical plant viruses and spherical animal viruses. Not only was the structure of picornaviruses such as polio, mengo, and rhino similar to each other and to plant viruses, the tertiary structure of the proteins is similar, suggesting some possible evolutionary ancestor. All the small icosahedral positive stranded RNA viruses studied so far are constructed on a common theme. This is a structure of 180 major building blocks of similar shape packed in an icosahedral array. The blocks are protein domains based on a eight strand anti parallel beta barrel. Most known domains of this kind contain 150-200 amino acids.

Helical Symmetry

The protomeres are not grouped in capsomeres, but are bound to each other so as to form a ribbon-like structure. This structure folds into a helix because the protomeres are thicker at one end than at the other. The diameter of the helical capsid is determined by characteristics of its protomeres, while the length of the nucleic acid it encloses determines its length. It is noteble that most proteins are non-symmetrical, however the virus is symmetrical; therefore have to find a way of arranging non-symmetrical units in a symmetrical fashion. Symmetry implies that each coat protein interacts with each other or neighbor in the same way. In the case of a helical virus such as TMV this is also true of interaction between protein and RNA. One of the simplest ways to do this is to arrange them in a circumference of a circle to form discs. The discs can then be stacked one on top of the other. The nucleic acid interacts with the proteins and forms a helical structure. "Linear" viral capsids have RNA genomes that are encased in a helix of identical protein subunits. The length of the nucleic acid determines the length of the helical viral nucleocapsid. Until 1960, the only known examples of virions with helical symmetry were those of plant viruses, the best-studied example being tobacco mosaic virus. T.M.V. is a good example of this.

T.M.V. virus is 180 A thick, 3000 A long: a central hole is 40 A running down the center. The capsid consists of 2,130 identical subunits, each with a mol.wt. of 17,000 daltons. The interaction of the N.A. and the capsids results in a helical turn every 16 1/3 subunit. RNA winds through this central core. Large advances have been made in X-ray crystallography of virus structure. Analysis of TMV has shown that the negative phosphates in the RNA chain are held in place by salt bridges to the positive chains of Arg-90 and Arg 92. The nature of theinteraction was not known. This virus undergoes self-assembly even in the absence of nucleic acid. Information for assembly is contained in the protein subunits. Infective viruses can self-assembly by mixing proteins and RNA. The key stable intermediate for self-assembly is a two-layered disk composed of 34 monomers of coat protein. TMV RNA contains a specific base sequence that initiates polymerization of the disks into virus particles. A disk binds preferentially to a region of RNA about 1000 bases from the 3' end. Since it has been found that both the 3' and the 5' ends of the RNA protrude from the same end of partially formed virus rods, it appears that a specific loop of RNA inserts into the hole in the disks. Binding of the disk to the RNA loop causes the conformation of the disk to change to a lock washer. Also phage M13.

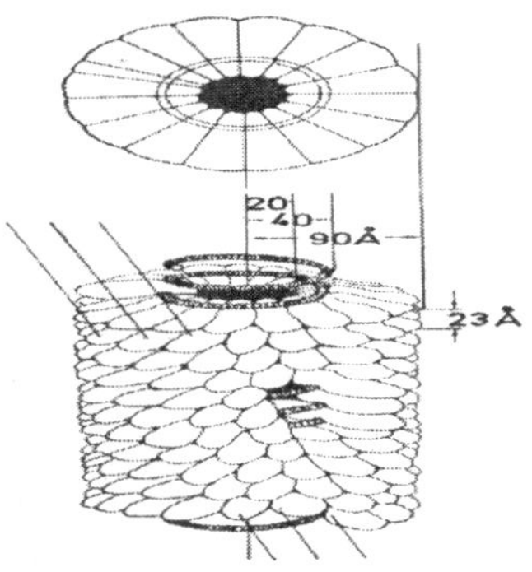

Fig. 2.11. Tobacco mosaic virus (TMV).

Among other viruses with similar structure are the myxoviruses, and the rhabdoviruses, except that there, the nucleocapsid is contained within an envelope. In the case of VSV, the RNA genome is about 11 kb long. The RNA genome and the basic nucleocapsid protein interact to form a helical structure, together with non-structural proteins make up the core of the virus. There are about 30-35 turns of nucleoprotein helix in the core which is about 180 nm long and 80nm in diameter Each capsid protein covers about 9 nucleotides. In common with most enveloped viruses the nucleocapsid is surrounded by an amorphous layer, which interacts with both the core and the envelope lipid layer. This is known as the matrix protein. This is usually the most abundant protein in the virus particle.

Complex Symmetry

Majority of viruses can be fitted into category of helical, icosahedral symmetry or enveloped viruses. However there are a few whose structure is more complex. Best example is the poxvirus group. Oval or brick shaped particle, 200-400 nm long. Particles extremely complex and contain more than 100 proteins. During replication two types of particles seen, extracellular forms that contain two membranes, and intracellular forms with one membrane. Outer surface of virus composed of lipid and proteins. This surrounds the core, which is a biconcave shape, and two lateral bodies of unknown function. Double stranded DNA wound around the core. Another example is the tailed phage of E. coli (Bacteriophage). T4 bacteriophage grows in the colon bacterium Escherichia coli. The bacteriophage is tadpole shaped, with head and tail regions.(Fig. 2.12). The head capsid is 95 nm x 65 nm and has the form of a prolate icosahedron. It is made up of about 2,000 similar subunits and is packed with circular double stranded DNA 500 um long. The head capsid may be described as consisting of two 10-faceted equatorial bands with a pyramidal vertex ateither end. The tail has helical symmetry. Thus the bacteriophage shows a combination of icosahedral symmetry and helical symmetry (banal symmetry). The tail consists of a core tube 80 A in diameter, through which DNA passes out, surrounded by a protein tail sheath. The sheath consists of 144 subunits arranged in 24 rings of 6 subunits each. The sheath is connected to a thin disc, called the collar, at the upper end and a base plate at the lower end. The base plate is hexagonal, and has a pin or spike a⁺ each corner. From each of the six corners is also given off a long , thin tail fibre, 1300 A long, which serves for the attachment of the bacteriophage to the host cell.

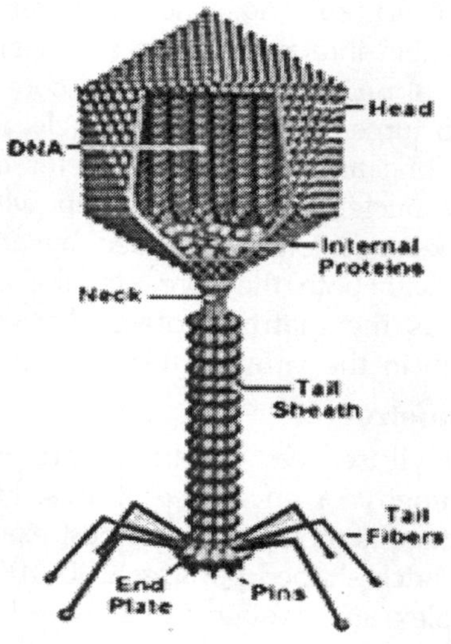

Fig. 2.12. Bacteriophage.

Another example of a complex virus is the insect baculovirus. This virus is important since it has become a prime expression vector for production of foreign proteins. They are natural pathogens of insects and are being investigated as agent for biological control of arthropods. These viruses contain 12-30 structural proteins, and consist of a rod like nucleocapsid containing 88-160kbp DNA. The nucleocapsid is surrounded by an envelope, outside of which there may be a crystalline protein matrix. If this outer shell is present the whole assembly is referred to as an occlusion body. The virus is said to be occluded. There are two general of occluded baculoviruses. Those with polyhedral occlusions 1000-1500 nm in diameter and may contain multiple nucleocapsids within the envelope e.g. nuclear polyhedrosis virus and those with ellipsoidal occlusions 200-500 nm e.g. granulosis virus. The function of these occlusion bodies is to confer resistance to environmental conditions. The occlusion bodies are alkali label and are removed in the pH environment of the insect gut. The occlusion bodies are known to be composed of many copies of a single protein of approx. 245 amino acids, the polyhedrin. This single gene product is hyper expressed late in infection by a very strong transcriptional promoter. This has been taken advantage of to form very effective expression vectors. An unusual arrangement found in the Gemini virus (a plant virus). The capsid is formed of two incomplete icosahedrons attached to each other, and it contains two different single stranded DNAs, both of which are required for infectivity. This virus consists of 110 protein monomers arranged in 22 morphological subunits.

Reproduction in Virus

Viruses can reproduce only within a host cell.Viral reproduction differs markedly from cellular reproduction, because viruses are obligate intracellular parasites, which can express their genes and reproduce only within a living cell. Each virus has a specific host range. Host range means limited number or range of host cells that a parasite can infect. Viruses recognize host cells by a complementary fit between external viral proteins and specific cell surface receptor sites. Some viruses have broad host ranges, which may include several species (e.g., swine flu and rabies).

Some viruses have host ranges so narrow that they can infect only one species (e.g., phages of E. coli), infect only a single tissue type of one species (e.g., human cold virus infects only cells of the upper respiratory tract; AIDS virus binds only to specific receptors on certain white blood cells).

There are many patterns of viral life cycles, but they all generally involve:

- Infecting the host cell with viral genome.
- Co-opting host cell's resources to:
  - Replicate the viral genome
  - Manufacture capsid protein
  - Assembling newly produced viral nucleic acid and capsomeres into the next generation of viruses

Viruses are remarkable "organisms" that economize dramatically in their use of genetic information as they have limited capacity (the volume of a viral capsid) for moving genetic information between cells of susceptible hosts. There are three possible patterns of viral genome replication:

1. DNA → DNA. If viral DNA is double-stranded, DNA replication resembles that of cellular DNA, and the virus uses DNA polymerase produced by the host.
2. RNA → RNA. Since host cells lack the enzyme to copy RNA, most RNA viruses contain a gene that codes for RNA replicase, an enzyme that uses viral RNA as a template to produce complementary RNA.
3. RNA → DNA → RNA. Some RNA viruses encode reverse transcriptase, an enzyme that transcribes DNA from an RNA template. Different kind of viruses replicate in different ways, depending on the structure of viral genome, but the process is basically the same.

The replication of viruses is performed through five major phases:

- Adsorption
- Penetration
- Biosynthesis
- Maturation
- Release

These steps differ somewhat among bacteriophages, animal viruses and plant viruses.

Replication in Bacteriophages

Bacteriophages (lytic cycle)

Tail fibers of the phage act like landing gear for the virus; they attach to specific receptors on the surface of bacterium. The phage tail sheath retracts like a syringe to inject the nucleic acid into cell. The capsid is left behind. Bacterial DNA becomes disrupted and phage DNA takes over cell functions. Viral DNA is copied to mRNA by host cell enzymes and new viruses are packaged in the host cell cytoplasm. The bacterial cell wall is subject to lysis and newly synthesized viruses are released.

Bacteriophages (lysogenic cycle)

Phages that are temperate live in a long-term stable relationship with their host rather than kill it outright. In this situation phage DNA integrates with the host cell DNA and forms a prophage. Viral DNA is replicated along with the bacterial DNA each time the cell divides. Viruses that carry genes for botulism and diphtheria toxins replicate in bacteria in this manner. From time to time the phage DNA can excise from the bacterium (it "senses" that it's time to find a new home) and undergo a lytic cycle. Viruses that transport genes from one bacterium to another do so by a process known as transduction.

Replication in Animal Viruses

Animal viruses have the phase of adsorption of particles on the surface of a cell by the certain receptors. At this level, a virus and a susceptible animal cells recognize each other. Host receptors are cell membrane proteins and glycoproteins. The capsid penetrates the cytoplasm either by fusion (as with enveloped viruses such as HIV) to the cell membrane or by endocytosis. Uncoating is required whereby the capsid proteins are removed enzymatically. Biosynthesis of viral particles occurs in the nucleus (DNA viruses) or the cytoplasm (RNA viruses). Many ssRNA viruses have positive (+) sense strand RNA; this functions directly as mRNA and can be translated into protein by host ribosomes.(example: Poliovirus). ssRNA viruses with a negative (-) antisense RNA genome must first transcribe a + sense strand that can function as mRNA before they can produce proteins (example: Rabies virus). The enzyme that produces a +RNA from a - strand is known as an RNA-dependent RNA polymerase. RNA viruses known as retroviruses, form a DNA copy from an RNA template using reverse transcriptase. This enzyme is an RNA-dependent DNA polymerase.

Before transcription occurs, the viral DNA must be integrated into the host cell DNA. This stage is known as a provirus, which, unlike a phage, does not excise from the host cell chromosome.

Replication in Plant Viruses

There is no recognition at that level with plant viruses, which is determined by differences in the build of a plant and animal cell. Plant viruses enter cells only through wounds, which are made mechanically or by vectors, or by deposition into an ovule, by an infected pollen grain. During the simplified replication of an RNA virus, its nucleic acid firstly gets rid of the protein coat and in that way the viral genome becomes free. By brining the viral genome into a cell, the genetic determinants, which code the synthesis of macromolecules specific to a virus, are also brought. The redirection of the cell's metabolism is performed by early proteins-enzymes (polymerase), which are coded by the viral genome. It is thought that one of the ways of the recognition of plant viruses and a susceptible host plant is exactly the synthesis of proteins, which are the components of polymerase. The ability of a virus to control a cell's metabolism depends on the nature of the virus and on the type of the cell. At the phase of biosynthesis viruses transmit the genetic information in different ways, and it is basic that specific mRNA must be transcribed (transcription) from the viral genome. When it is once achieved, viruses then translate (translation) mRNA into structural and non-structural proteins by using cellular components and mechanisms. After this phase, proteins and nucleic acid are self-assembled into viral particles, which are further able to infect.

Replication of Nucleic Acid

Replication of viral nucleic acid is a complex and variable process. The specific process depends on the nucleic acid type.

DNA virus replication

With the exception of the poxviruses, all DNA viruses replicate in the nucleus. In some cases one of the DNA strands is transcribed (in others both strands of a small part of the DNA may be transcribed) into specific mRNA, which in turn is translated to synthesize virus-specific proteins such as tumor antigen and enzymes necessary for biosynthesis of virus DNA. This period encompasses the early virus functions. Host cell DNA synthesis is temporarily elevated and is then suppressed as the cell shifts over to the manufacture of viral DNA. As the viral DNA continues to be transcribed, late virus functions become apparent. Messenger RNA transcribed during the later phase of infection migrates to the cytoplasm and is translated. Proteins for virus capsids are synthesized and are transported to the nucleus to be incorporated into the complete virion. Assembly of the protein subunits around the viral DNA results in the formation of complete virions, which are released after cell lysis. The single-stranded DNA viruses first form a double stranded DNA, utilizing a host DNA-dependent DNA polymerase. They then undergo a typical replication cycle.

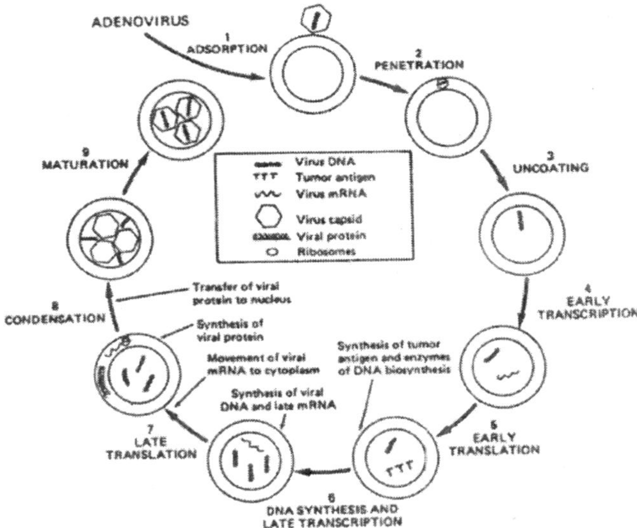

Fig. 2.13. Steps in the replication of adenovirus contained DNA in its genome.

RNA virus replication

With the exception of the orthomyxoviruses and retroviruses, all RNA viruses replicate in the cytoplasm of the host cell. The exact process varies with the species of virus. The single-stranded RNA that is released after uncoating will act as either the mRNA to synthesize viral-coded proteins; or a template to synthesize mRNA; or a template to synthesize double stranded RNA, which is then used as a template to synthesize mRNA; or a template to synthesize double-stranded DNA, which is then utilized as a template to synthesize mRNA. This latter process occurs only with the retroviruses (oncornaviruses). The replication of poliovirus, which contains a single-stranded RNA as its genome, provides a useful example. All of the steps are independent of host DNA and occur in the cell cytoplasm. Polioviruses absorb to cells at specific cell receptor sites, losing in the process one virus polypeptide. The sites are specific for virus coat-cell interactions. After attachment, the virus particles are taken into the cell by viropexis (similar to pinocytosis), and the viral RNA is uncoated. The single-stranded RNA then serves as its own messenger RNA. This messenger RNA is translated, resulting in the formation of an RNA-dependent RNA polymerase that catalyzes the production of a replication intermediate (RI), a partially double-stranded molecule consisting of a complete RNA strand and numerous partially completed strands. At the same time, inhibitors of cellular RNA and protein synthesis are produced. Synthesis of (+) and (-) strands of RNA occurs by similar mechanisms. The RI consists of one complete (-) strand and many small pieces of newly synthesized (+) strand RNA. The replicative form (RF) consists of two complete RNA strands, one (+) and one (-). The single (+) strand RNA is made in large amounts and may perform any one of three functions: (i) serve as messenger RNA for synthesis of structural proteins; (ii) serve as template for continued RNA replication; or (iii) become encapsulated, resulting in mature progeny virions. The synthesis of viral capsid proteins is initiated at about the same time as RNA synthesis. The entire poliovirus genome acts as its own mRNA, forming a polysome of approximately 350S, and is translated to form a single large polypeptide that is subsequently cleaved to produce the various viral capsid polypeptides. Thus, the poliovirus genome serves as a polycistronic messenger molecule. Poliovirus contains four polypeptides.

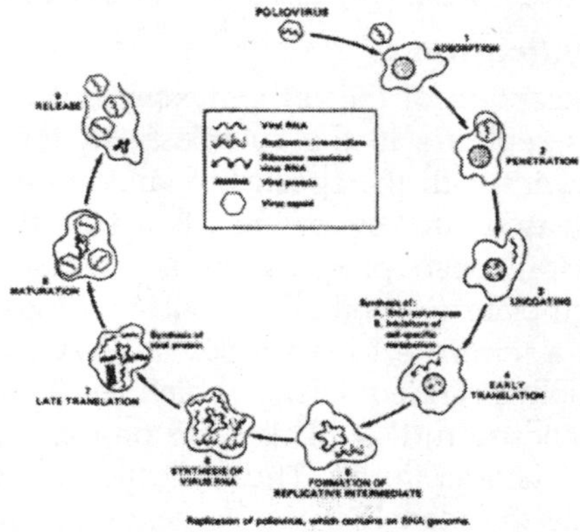

Fig. 2.14. Replication in Polio virus.

Maturation and Release

Naked viruses

Maturation consists of two main processes: the assembly of the capsid, and its association with the nucleic acid. Maturation occurs at the site of nucleic acid replication. After they are assembled into mature viruses, naked virions may become concentrated in large numbers at the site of maturation, forming inclusion bodies. Naked virions are released in different ways, which depend on the virus and the cell type. Generally, RNA-containing naked viruses are released rapidly after maturation and there is little intracellular accumulation; therefore, these viruses do not form predominant inclusion bodies. On the other hand, DNA-containing naked icosahedral viruses that mature in the nucleus do not reach the cell surface as rapidly, and are released when the cells undergo autolysis or in some cases are extruded without lysis. In either case they tend to accumulate within the infected cells over a long period of time. Thus, they generally produce highly visible inclusion bodies.

Enveloped viruses

In the maturation of enveloped viruses, a capsid must first be assembled around the nucleic acid to form the nucleocapsid, which is then surrounded by the envelope. During the assembly of the nucleocapsid, virus-coded envelope proteins are also synthesized. These migrate to the plasma membrane (if assembly occurs in the cytoplasm) or to the nuclear membrane (if assembly occurs in the nucleus) and become incorporated into that membrane. Envelopes are formed around the nucleocapsids by budding of cellular membranes. Enveloped viruses will have an antigenic mosaicism characteristic of the virus and the host cell. The budding process with the results slowly and continuously releases viruses that: (a) the cell is not lysed; and (b) little intracellular accumulation of virus occurs; and (c) inclusion bodies are not as evident as with naked viruses.

Complex viruses

These viruses, of which the poxvirus is a good example, begin the maturation process by forming multilayered membranes around the DNA. These layers differentiate into two membranes: The inner one contains the characteristic nucleoid, while the external one acquires the characteristic pattern of the surface of the virion.

3

CLASSIFICATION OF VIRUS

THE INTERNATIONAL CODE OF VIRUS CLASSIFICATION AND NOMENCLATURE

Statutory basis for the International Committee on Taxonomy of Viruses (ICTV)

1. The International Committee on Taxonomy of Viruses (ICTV) is a committee of the Virology Division of the International Union of Microbiological Societies. ICTV activities are governed by Statutes agreed with the Virology Division.
2. The Statutes define the objectives of the ICTV. These are:
   (i) to develop an internationally agreed taxonomy for viruses
   (ii) to develop internationally agreed names for virus taxa, including species and subviral agents.
   (iii) to communicate taxonomic decisions to the international community of virologists.
   (iv) to maintain an Index of virus names.
3. The Statutes also state that classification and nomenclature will be subject to Rules set out in an International Code.

Principles of Nomenclature

1. The essential principles of virus nomenclature are:-
   (i) aim for stability.
   (ii) avoid or reject the use of names which might cause error or confusion.
   (iii) avoid the unnecessary creation of names.

2. Nomenclature of viruses and sub-viral agents is independent of other biological nomenclature. Virus and virus taxon nomenclature are recognised to have the status of exceptions in the proposed International Code of Bionomenclature (BioCode).
3. The primary purpose of naming a taxon is to supply a means of referring to the taxon, rather than to indicate the characters or the history of the taxon.
4. The application of names of taxa is determined, explicitly or implicitly, by means of nomenclatural types.
5. The name of a taxon has no status until it has been approved by ICTV.

Rules of Classification and Nomenclature (General Rules)

The universal scheme

1. Virus classification and nomenclature shall be international and shall be universally applied to all viruses.
2. The universal virus classification system shall employ the hierarchical levels of Order, Family, Subfamily, Genus, and Species.

   Comments: It is not obligatory to use all levels of the taxonomic hierarchy. The primary classification is of viruses into species. Most species are classified into genera and most genera are classified into families. Species not assigned to a genus will be "unassigned" in a family (see Rule 3.6) and genera not classified in families have the status of "unassigned" (sometimes referred to as "floating"). Some families are classified together into Orders, but for many, the family is the highest level taxon in use. Also, families are not necessarily divided into subfamilies. This taxon is to be used only when it is needed to solve a complex hierarchical problem.

   Contrasting examples of full classifications of some negative strand RNA viruses are: (1) species Mumps virus; genus Rubulavirus; subfamily Paramyxovirinae; family Paramyxoviridae; order Mononegavirales, and (2) species Rice stripe virus; genus Tenuivirus.

Scope of the classification

3. The ICTV is not responsible for classification and nomenclature of virus taxa below the rank of species. The classification and

naming of serotypes, genotypes, strains, variants and isolates of virus species is the responsibility of acknowledged international specialist groups.

Comments: Particular virus isolates may be regarded as strains, variants, clusters or other subspecific entities that, together with other entities, constitute a species. Classification of such isolates is not the responsibility of the ICTV but is the responsibility of international specialty groups. It is the responsibility of ICTV Study Groups to decide if an isolate or a group of isolates should constitute a species.

Deciding the names of serotypes, genotypes, strains, variants or isolates of virus species is not the responsibility of the ICTV. However, it is recommended that new names not be the same as, or closely similar to, names already in use (Rule 3.14 for taxa). When a particular virus isolate is designated to represent a species, the decision as to which name will be adopted for the species for formal taxonomic purposes will be the responsibility of the ICTV, initially of a particular Study Group working on behalf of the ICTV. The Study Group will be expected to consult widely so as to ensure the acceptability of names, subject to the Rules in the Code. The policy of the ICTV is that as far as is possible, taxonomic and nomenclatural decisions should reflect the majority view of the appropriate virological constituency.

4. Artificially created viruses and laboratory hybrid viruses will not be given taxonomic consideration. Their classification will be the responsibility of acknowledged international specialist groups.

   Comments: Naturally occurring isolates that have genomes formed from parts of the genomes of different strains of a virus, either by recombination between the genome nucleic acids or by re-assortment of separate genome parts, will be classified either as species or subspecific entities in the same way that other isolates are classified. Neither artificial variants made by recombination or re-assortment, nor mutant viruses, are subject to the Rules in the Code.

Limitations

5. Taxa will be established only when representative member viruses are sufficiently well characterized and described in the

published literature so as to allow them to be identified unambiguously and the taxon to be distinguished from other similar taxa.

6. When it is uncertain how to classify a species into a genus but its classification in a family is clear, it will be classified as an unassigned species of that family.

   Comments: A species can be classified as an unassigned member of a family when no genus has been devised. For example, Bimbo virus is a rhabdovirus of vertebrates but is not a member of any of the currently recognised genera in the family Rhabdoviridae. Likewise, Groundnut rosette assistor virus resembles viruses in the family Luteoviridae but is not classified in any of the genera in that family. These viruses are each classified as an unassigned member of their respective families.

7. Names will only be accepted if they are linked to taxa at the hierarchical levels described in Rule 3.2 and which have been approved by the ICTV.

   Comments: Taxa above the rank of species must be approved before a name is assigned to them. Proposals for the creation of taxa shall be accompanied by proposals for names. A decision to create a taxon can thus be followed immediately by a decision about the name for the taxon. Species will be approved together with their names as a single taxonomic act.

   The following example is of a proposal concerning an imaginary virus with the vernacular name of "beta gamma virus" that is related to another virus, "alpha beta virus".

   Proposal 1. Approve Beta gamma virus as a species containing strains known as "beta gamma virus" and "alpha beta virus".

   Proposal 2. Create a genus to contain species resembling Beta gamma virus.

   Proposal 3. Name the genus created by Proposal 2, Betavirus.

   Proposal 4. Nominate Beta gamma virus as the type species of the genus Betavirus.

   Proposal 5. Create a family to contain genus Betavirus and similar genera.

Proposal 6. Name the family created by Proposal 5, Betaviridae.

Proposal 7. Assign species X, Y and Z to genus Betavirus (such a proposal should include a listing of the parameters for discriminating between species in the genus Betavirus).

Rules about naming Taxa

Status of Names

8. Names proposed for taxa are "valid names" if they conform to the Rules set out in the Code and they pertain to established taxa. Valid names are "accepted names" if they are recorded as approved International Names in the 6th ICTV Report or become "accepted names" by an ICTV vote of approval for a taxonomic proposal.

   Comments: A valid name is one that has been published, one that is associated with descriptive material, and one that is acceptable in that it conforms to the Rules in the Code. Accepted names will be kept in an "Index" by the ICTV.

9. Existing names of taxa and viruses shall be retained whenever feasible.

   Comment: A stable nomenclature is one of the principal aims of taxonomy and therefore changes to names that have been accepted will only be considered in exceptional circumstances, and then only because of serious conflict with the Rules.

10. The rule of priority in naming taxa and viruses shall not be observed.

    Comments: The earlier of candidate names for a taxon may be chosen as a convenience to virologists, but the Rule ensures that it is not possible to invalidate a name in current use by claiming priority for an older name that has been superceded.

11. No person's name shall be used when devising names for new taxa.

    Comments: New taxon names shall not be made by adopting a person's name, by adding a formal ending to a person's name or by using part of a person's name to create a stem for a name. When existing names of species incorporate a person's name (for example, Shope papilloma virus) continued usage of this name, in agreement with Rule 3 and 9, is in general preferable to the creation of a new name.

12. Names for taxa shall be easy to use and easy to remember. Euphonious names are preferred.

    Comments: In general, short names are desirable and the number of syllables should be kept to a minimum.

13. Subscripts, superscripts, hyphens, oblique bars and Greek letters may not be used in devising new names.

    Comments: The Rule is intended to make text unambiguous and easy to manipulate and its application should often make names more pronouncable, in agreement with Rule 12. Existing names of some species violate this Rule (e.g. coliphage l), but international names of genera and families do not.

14. New names shall not duplicate approved names. New names shall be chosen such that they are not closely similar to names that are in use currently or have been in use in the recent past.

    Comments: The name selected for a new taxon should not sound indistinguishable from the name of another taxon at any rank or from any taxon. For example, the existence of the genus Iridovirus means that forms of new name such as "irodovirus" or "iridivirus" are unacceptable as they are too easily confused with an approved name. Confusion can also be between species and genus names as both end in "virus". Thus, for example, the name selected for a genus typified by a species "Omega virus" would not be named "Omegavirus" because species and genus would then be too readily confused.

15. Sigla may be accepted as names of taxa, provided that they are meaningful to virologists in the field, normally as represented by Study Groups.

    Comments: Sigla are names comprising letters and/or letter combinations taken from words in a compound term. The name of the genus Comovirus has the sigla stem "Co-" from cowpea and "-mo-" from mosaic; the name of the genus Reovirus has the sigla stem "R" from "Respiratory, "e" from "enteric" and "o" from "orphan".

Decision making

16. In the event of more than one candidate name being proposed, the relevant Subcommittee will make a recommendation to the Executive Committee of the ICTV,

which will then decide among the candidates as to which to recommend to ICTV for acceptance.

Comments: When there is more than one candidate name for the same taxon, the decision as to which will be accepted shall be made on the basis of the Code and, if necessary, thereafter on the basis of likely acceptability to the majority of virologists.

17. If no suitable name is proposed for a taxon, the taxon may be approved and the name be left undecided until the adoption of an acceptable international name, when one is proposed to and accepted by ICTV.

    Comments: When genera have not been named this is indicated by quotation marks. For example, Soybean chlorotic mottle virus is a species in an as yet un-named genus in the family Caulimoviridae. Until the genus is named, it is designated as "Soybean chlorotic mottle-like viruses". This designation is regarded as temporary as it is an inconvenience to most virologists.

18. New names shall be selected such that they, or parts of them, do not convey a meaning for the taxon which would either (1) seem to exclude viruses which lack the character described by the name but which are members of the taxon being named, or (2) seem to exclude viruses which are as yet undescribed but which might belong to the taxon being named, or (3) appear to include within the taxon viruses which are members of different taxa.

19. New names shall be chosen with due regard to national and/or local sensitivities. When names are universally used by virologists in published work, these or derivatives shall be the preferred basis for creating names, irrespective of national origin.

Procedures for naming taxa

20. Proposals for new names, name changes, establishment of taxa and taxonomic placement of taxa shall be submitted to the Executive Committee of the ICTV in the form of taxonomic proposals. All relevant ICTV subcommittees and study groups will be consulted prior to a decision being taken.

    Comments: For example, taxonomic proposals concerned with the family Partitiviridae would be considered by the Fungal Virus Subcommittee and one of its Study Groups but because some genera in the family contain viruses of plants, proposals affecting the family would also be considered by the Plant Virus Subcommittee.

Rules about Species

Definition of a virus species

21. A virus species is defined as a polythetic class of viruses that constitutes a replicating lineage and occupies a particular ecological niche.

Definition of "tentative" status

22. When an ICTV Subcommittee is uncertain about the taxonomic status of a new species or about the assignment of the new species to an established genus, the new species will be listed as a tentative species in the appropriate genus or family. Names of tentative species, as of taxa generally (Rule 14), shall not duplicate approved names and shall be chosen such that they are not closely similar to names that are in use currently, names that have been in use in the recent past, or names of definitive species.

    Comments: Species classified as tentative are candidates for taxonomic decision by the appropriate Study Groups to resolve their tentative status.

Construction of a name

23. A species name shall consist of as few words as practicable but shall not consist only of a host name and the word "virus".

    Comments: The styles used when virus names are devised differ according to the traditions of the particular fields of virology. For example, plant virus names are usually constructed as host + symptom + "virus" (e.g. tobacco necrosis virus) whereas, in contrast, viruses in the family Bunyaviridae are usually named after the location at which the virus was found + "virus" (e.g. Bunyamwera virus).

24. A species name must provide an appropriately unambiguous identification of the species.

25. Numbers, letters, or combinations thereof may be used as species epithets where such numbers and letters are already widely used. However, newly designated serial numbers, letters or combinations thereof are not acceptable alone as species epithets. If a number or letter series is in existence it may be continued.

Rules about Genera

26. A genus is a group of species sharing certain common characters.
27. A genus name shall be a single word ending in virus.
28. Approval of a new genus must be accompanied by the approval of a type species.

Rules about Subfamilies

29. A subfamily is a group of genera sharing certain common characters. The taxon shall be used only when it is needed to solve a complex hierarchical problem.
30. A subfamily name shall be a single word ending in virinae.

Rules about Families

31. A family is a group of genera (whether or not these are organized into subfamilies) sharing certain common characters.
32. A family name shall be a single word ending in viridae.

Rules about Orders

33. An order is a group of families sharing certain common characters.
34. An order name shall be a single word ending in virales.

Rules about Sub-viral Agents

Viroids

35. Rules concerned with the classification of viruses shall also apply to the classification of viroids.
36. The formal endings for taxa of viroids are the word "viroid" for species, the suffix "-viroid" for genera, the suffix "-viroinae" for sub-families (should this taxon be needed) and "-viroidae" for families.

    Comments: For example, the species Potato spindle tuber viroid is classified in genus Pospiviroid, and the family Pospiviroidae.

Other sub-viral Agents

37. Retrotransposons are considered to be viruses in classification and nomenclature
38. Satellites and prions are not classified as viruses but are assigned an arbitrary classification as seems useful to workers in the particular fields.

Rules for Orthography

39. In formal taxonomic usage, the accepted names of virus Orders, Families, Subfamilies, and Genera are printed in italics and the first letters of the names are capitalized.

    Comments: See Rule 8 for the definition of an "accepted" name.

40. Species names are printed in italics and have the first letter of the first word capitalized. Other words are not capitalized unless they are proper nouns, or parts of proper nouns.

    Comments: When used formally, as labels for taxonomic entities, the names "Tobacco mosaic virus" and "Murray River encephalitis virus" are in the correct form and typographical style. Examples of incorrect forms are Aspergillus niger virus S (not italic), Murray river encephalitis virus (River is a proper noun) or tobacco mosaic virus (not capitalized or italic).

    Taxa are abstractions and thus when their names are used formally, these are written distinctively using italicization and capitalization. In other senses, such as an adjectival form (e.g. the tobacco mosaic virus polymerase) italics and capital initial letters are not needed. Equally, these are not needed when referring to physical entities such as virions (e, g. a preparation or a micrograph of tobacco mosaic virus).

    This Rule was introduced in 1998 and is in contradistinction to Rules in the Code published in the 6th ICTV Report.

41. In formal usage, the name of the taxon shall precede the term for the taxonomic unit.

    Comments: For example, the correct formal descriptions of various taxa are the family Herpesviridae the genus Morbillivirus, the genus Rhinovirus, the species Tobacco necrosis virus, and so on.

The International Committee on Taxonomy of Viruses (ICTV)

Against this background, in 1966 the International Committee on Nomenclature of Viruses (ICNV) was established at the International Congress of Microbiology in Moscow. At that time, virologists already sensed a need for a single, universal taxonomic scheme. There was little dispute that the hundreds of viruses being isolated from humans, animals, plants, invertebrates, and bacteria, should be classified in a single system, and that this system should separate the viruses from all other biological entities. Nevertheless, there was much dispute over the taxonomic system to be used. Lwoff, Horne and Tournier (1962) argued for the adoption of an all-embracing scheme for the classification of viruses into subphyla, classes, orders, suborders, and families. Descending hierarchical divisions were to be based, arbitrarily and monothetically, upon nucleic acid type, capsid symmetry, presence or absence of an envelope, etc. Opposition to this scheme was based upon its arbitrariness in deciding the relative importance of virion characteristics to be used and upon the argument that not enough was known about the characteristics of most viruses to warrant an elaborate hierarchy. An alternative proposal was set forth in 1966 by Gibbs et al. (1966); in this system, divisions were based upon multiple criteria (polythetic criteria). The system was illustrated by the use of "cryptograms" (coded notations of eight virus characters). These early efforts succeeded well in stimulating interest in the development of the universal taxonomy system that evolved in the 1970s and has been built upon ever since (Wildy, 1971; Matthews, 1983).

In the universal scheme developed by the ICTV, virion characteristics are considered and weighted as criteria for making divisions into families, in some cases subfamilies, and genera (until recently, the scheme did not use any hierarchical level higher than that of family, but now one order, the order Mononegavirales, has been approved). In each case, the relative hierarchy and weight assigned to each characteristic used in defining taxa is set arbitrarily and is still influenced by prejudgments of relationships that "we would like to believe (from an evolutionary standpoint), but are unable to prove" (Fenner, 1974). As the species taxon has been developed in the 1990s, it has become clearer that families and genera might best be defined monothetically (or by just a few characters), but species are better-defined polythetically (Van Regenmortel, 1990).

At its meeting in Mexico City in 1970, the ICTV approved the first two families and 24 floating genera (Wildy, 1971; Matthews, 1983). At that time, 16 plant virus groups were also designated (Harrison, et al., 1966). Since then, the ICTV has published five Reports entitled The Classification and Nomenclature of Viruses (Wildy, 1971; Fenner, 1976; Matthews, 1979; Matthews, 1982; Francki et al., 1991). Additionally, the Study groups of the

ICTV published over the years detailed descriptions of the characteristics of the member viruses of many taxa (e.g., Melnick et al., 1974; Pfau et al., 1974; Dowdle et al., 1975; Cooper et al., 1978; Kingsbury et al., 1978; Porterfield et al., 1978; Brown et al., 1979; Schaffer et al., 1980; Bishop et al., 1980; Roizman et al., 1982, 1992; Kiley et al., 1982; Wigand et al., 1982; Gust et al., 1983; Siddell et al., 1983; Siegl et al., 1985; Westaway et al., 1985; Gust et al., 1986; Brown, 1986). In the Sixth Report of the ICTV, records a universal taxonomy scheme comprising one order, 71 families, 9 subfamilies, and 164 genera, including 24 floating genera, and more than 3,600 virus species. The system still contains hundreds of unassigned viruses, largely because of a lack of data.

The Advance of Phylogenetic Taxonomy and the Use of Higher Taxa

Until recently, one of the rules of virus taxonomy stated that the system was not meant to imply any phylogenetic relationships. Since viruses leave no fossils (except perhaps within arthropods and other creatures embedded in amber!), it was presumed that there never would be enough evidence to prove whether or not different taxa had common evolutionary roots. In fact, since the prevailing concept was that each kind of virus was derived separately from its host, it was considered foolish to consider any idea of a single evolutionary "tree" for the viruses (Goldbach, 1986, Gibbs, 1987). The generally very different morphological and physicochemical characteristics of the member viruses of many different taxa supported this view.

Now, as genome sequencing of many viruses, and many organisms from archaebacteria to humans, is revealing many conserved functional or "fossil" domains of ancient lineage, for the first time the archeology of the viruses is being explored from the perspective of data, not just "armchair theory." Now, it is clear many viruses have gained some functional genes from their hosts (and hosts have gained some genes from viruses) and have gained other genes from other viruses - that is, viral genomes seem to represent more-or-less ancient "grab-bags" of genes, fine-tuned by the Darwinian forces of selection into replicative machines with extraordinary functional economy. It is also known that the genomes of viruses in different families, in most cases, are extremely different from each other, but in some cases seemingly unrelated viruses (and taxa) are similar - similar in gene order and arrangement, fine points of strategy of replication, and even in conserved sequence domains encoding similarly functioning proteins. Overall, the differences between most taxa are so great that it still seems foolish to think of building a monophylogenous "tree" uniting all the viruses. On the other hand, the unexpected similarities have prompted some consideration of a partial phylogenetic taxonomy (Goldbach, 1986, 1987; Gibbs, 1987; Goldbach and Wellink, 1988; Kingsbury, 1988; Morse, 1993).

As these evidences of phylogenetic relationships between families have been studied, there has been a wish to reflect these relationships in the universal taxonomic scheme. There has been no wish to combine families exhibiting distant phylogenetic relationships and thereby have them lose their practical identities. As one virologist stated: "the family is the fixed point, the benchmark, in virus taxonomy, so let's not do anything to change this". Instead, there has been increasing interest in capturing these relationships by uniting distantly related families in higher taxa, namely orders. The order Mononegavirales, comprising the families Paramyxoviridae, Rhabdoviridae and Filoviridae, was formed in recognition that the member viruses had common sequences in their nucleocapsid genes and similar gene arrangements and gene products (Pringle, 1991). At this point, ICTV is committed to reserving the hierarchical level of order solely for recognizing phylogenetic relationships.

There is a further occasion for considering the grouping of families together; this involves many of the positive sense, single-stranded RNA viruses. Similarities in genome organization, gene arrangements and sequence similarities in particular domains among viruses in several taxa, representing diverse vertebrate, invertebrate, plant and bacterial viruses, have been studied for the past 10 years. Kamer and Argos (1984) first aligned RNA-dependent RNA polymerase gene sequences of several plant, animal and bacterial viruses. In 1986, Goldbach greatly broadened this approach and used the data to explore the possible paths of evolution of the many positive sense RNA viruses. He proposed the formation of several "supergroups" to formalize the recognition of similarities. Gibbs (1987) and Strauss, Strauss and Levine (1988, 1991) have explored the possible mechanisms underpinning such evolutionary relationships. Presently work is centered on assessment of many characteristics of the RNA viruses: (1) genome organization and gene order; (2) presence of a 5'-terminal covalently-linked polypeptide, a 3'-terminal poly (A) tract, a 5'-terminal cap; (3) presence of subgenomic RNA; (4) polyprotein processing and enzymology; etc. At the heart of the matter is the assessment of conserved sequences in genes encoding the RNA-dependent RNA polymerases, helicases, and proteases. These characteristics are being used as the basis for several different ideas for constructing higher taxa (Koonin and Gorbalenya, 1989; Koonin, 1991; Mahy, 1991; Goldbach et al., 1991; Strauss, Strauss and Levine, 1991; Koonin and Dolja, 1993; Ward, 1994).

In their most recent proposals, Goldbach and de Haan (1993), Koonin and Dolja (1993) and Ward (1994) have proposed three major clusters of positive sense, single-stranded RNA viruses, one for the "picorna-like viruses," one for the "toga-like viruses," and one for the "flavi-like viruses" (although Goldbach calls attention to the major differences among the viruses in the latter cluster). Goldbach and de Haan (1993) continue to use the terms "supergroups" and "superfamilies" for these clusters, but Koonin and Dolja (1993) and Ward (1994) have taken this a step further in using the taxonomic hierarchical levels of class and order to denote the same groupings. It was thought for a time that the approach could also been extended to the double-stranded RNA viruses, but recent evidence suggests a polyphyletic origin of double-stranded RNA viruses from different groups of positive sense RNA viruses (Bruenn, 1991; Koonin, 1992). This matter will be debated by the ICTV over the next few years, but already it is clear that there is no general wish by most virologists to abandon the present system which is based upon assessment and weighting of multiple virion characteristics. As one virologist has stated that he is alarmed by the idea of erecting higher taxa upon a scheme that assumes that the polymerase is the virus." Clearly, there will be interest in melding phylogenetic considerations and traditional approaches into a unifying system.

Within the subject of phylogenetic taxonomy, one of the most interesting debates centers on which characteristics of an organism are most ancient, which are most recent, which are most stable and which are most changing. With increasing knowledge of which characteristics are conserved through evolutionary divergence, arbitrary cladistic taxonomy seemingly must be melded with phylogenetic taxonomy. Many virologists have over the years considered that virion structural elements were most ancient; after all, cumulative mutations in icosahedral capsids could only lead to lethal instability. Similarly, many virologists have considered that viral genome expression strategies were most ancient; again, even if a virus figured out how to reinvent its capsid, changes in integrated multigenic replication steps would certainly be lethal. At present, however, the finding of conserved sequence domains in polymerases, helicases and proteases, but not in structural or other genes of the positive sense, single-stranded RNA viruses, suggests that the whole subject must be revisited. Of course, horizontal gene transfer between viruses and dynamic gene acquisition from host cells adds to the sense of phylogenetic complexity. It is unfortunate that viruses have such small genomes, so that there will not be an opportunity to find confirmatory evidences of relationships by analyzing additional genes.

The ICTV Database (ICTVdB)

The number of viruses occupying geographic and/or host niches as pathogens or silent passengers of humans, animals, plants, invertebrates, protozoa, fungi, and bacteria is very large. The ICTV lists are increasing regularly as they search in new niches and as the sensitivity and specificity of their techniques for detection get better and better. Today, the ICTV recognizes more than 3,600 virus species. Specialty groups keep track of far more viruses, virus strains, and subtypes, each having particular health or economic distinction and importance. It has been estimated that more that 30,000 viruses, virus strains, and subtypes are being tracked in various specialty laboratories, reference centers, and culture collections communicating with the WHO, FAO, and other international agencies. Further, the development of the viral quasispecies concept, with its prediction of rapid evolution of variants that may become fixed in nature as new species, portends future needs to track even more viral entities (Holland, et al., 1982; Zimmern, 1988; Holland, de la Torre and Steinhauer 1992; Dolja and Carrington, 1992; Eigen, 1993). It has been estimated that to describe a virus comprehensively, approximately 500-1,000 characters must be determined (Atherton, Holmes and Jobbins, 1983; Boswell, et al., 1986). This means that to comprehensively describe all the known viruses of the world, they must "fill in the blanks" for 3 to 21 million data points (of course, many of the data points are the same when entering related viruses). This situation is even more complex; as they add more and more genome sequence information, their data systems will become truly enormous.

A major goal of the ICTV is to design, build and make available to all virologists, worldwide, a universal virus database, the ICTVdB. This database will encompass data that are now used in developing and managing the universal system for virus taxonomy. The ICTVdB will first describe viruses down to the species level, in keeping with the level of responsibility of the ICTV, but it will then go further, interfacing with the databases of international specialty groups which are cataloguing data down to subspecies, strain, variant and isolate levels, that is, levels important in medicine, agriculture, and other scholarly fields. ICTV's goal is to design the database to feed directly into user friendly programs that will be directly accessible to users (Pankhurst and Aitchison, 1975). Particular products, tailored to particular users, will be compressed to fit into equipment and software that can be readily accessed around the world.

There are several virus databases in operation in the world which will be integrated into the ICTVdB, including:

1. The plant virus database operated at the Australian National University in Canberra, Australia (the VIDE Project; AJ Gibbs, personal communication, 1994).
2. The veterinary virus database operated by the CSIRO/ Australian Animal Health Laboratory in Geelong, Australia (the VIREF Project; 1994).
3. An arbovirus database operated for the American Committee on Arthropod-borne Viruses (ACAV) by the Centers for Disease Control in Ft. Collins, Colorado, USA, 1994.

Additionally, a human virus/human disease database, in planning stage in the Division of Viral and Rickettsial Diseases, Centers for Disease Control in Atlanta, Georgia, USA, will be integrated into the ICTVdB (B.W.J. Mahy, 1994). The most advanced of these databases is the VIDE (Virus Identification Data Exchange) project on plant viruses (Dallwitz, 1974, 1980; Boswell et al., 1983, 1986; Dallwitz and Paine, 1986; Partridge Dallwitz and Watson, 1988). This project, centered in Canberra, Australia, interfaces with Horticulture Research International (Littlehampton,

U.K.) and C.A.B. International (Wallingford, U.K.), and involves a worldwide network of more than 200 collaborating plant virologists (Buchen-Osmond et al., 1988, 1993; Brunt, Crabtree and Gibbs, 1990, 1992). The VIDE database now contains 569 characters for more than 890 plant virus species in 55 genera (A.J. Gibbs, personal communication, 1994). Using these databases as a foundation, the ICTV has laid out a plan to develop the universal virus database, the ICTVdB, over the next 10 years.

THE UNIVERSAL SYSTEM OF VIRUS TAXONOMY

Now, a days there is a sense that a significant fraction of all existing viruses of humans, domestic animals and economically important plants have already been isolated and entered into the taxonomic system. This sense is based upon the infrequency in recent years of discoveries of viruses that do not fit into present taxa. Of course, this sense does not extend to the viruses infecting the myriad of other species populating the Earth. This present sense of the diversity of the viruses, however imperfect, does point once again to the need for a universal, usable taxonomic system - a system to keep track of the large numbers of different viruses being isolated and studied throughout the world, a system to tie viral characteristics to virus names. The present universal system of virus taxonomy is useful and usable. It is set arbitrarily at hierarchical levels of order, family, subfamily, genus, and species. Lower hierarchical levels, such as subspecies, strain, variant, etc., are established by international specialty groups and by culture collections.

Virus Orders

Virus orders represent groupings of families of viruses that share common characteristics and are distinct from other orders and families. Virus orders are designated by names with the suffix -virales. To date, one order has been approved by the ICTV, the order Mononegavirales, comprising the families Paramyxoviridae, Rhabdoviridae and Filoviridae. It is ICTV's intention to move slowly in the approval of orders, limiting use to those instances where there is good evidence of phylogenetic relationship among the viruses of member families.

Virus Families and Subfamilies

Virus families represent groupings of genera of viruses that share common characteristics and are distinct from the member viruses of other families. Virus families are designated by names with the suffix -viridae. Despite concerns about the arbitrariness of early criteria for creating these taxa, most of the original families have stood the test of time and are still intact. This level in the taxonomic hierarchy now seems stable, and, indeed, is the benchmark of the entire universal taxonomy system. Most of the families of viruses have distinct virion morphology, genome structure, and/or strategies of replication, indicating phylogenetic independence or great phylogenetic separation. At the same time, the virus family is being recognized as a taxon uniting viruses with a common, even if distant phylogeny. In four families, namely the families Poxviridae, Herpesviridae, Parvoviridae, and Paramyxoviridae, subfamilies have been introduced to allow for a more complex hierarchy of taxa, in keeping with the apparent intrinsic complexity of the relationships among member viruses. Subfamilies are designated by terms with the suffix -virinae.

Virus Genera

Virus genera represent groupings of species of viruses that share common characteristics and are distinct from the member viruses of other genera. Virus genera are designated by terms with the suffix -virus. This level in the hierarchy of taxa also seems stable and in many cases may be considered a benchmark for setting definitions of other taxa, especially species. The criteria used for creating genera differ from family to family. As more viruses are discovered and studied, there is pressure in many families to use smaller and smaller genetic, structural or other differences to create new genera. Since evidence of common phylogeny has entered the definition of many families, it is logical that even more such evidence will become the basis for defining genera. In fact, it might be said that in the future genera will not stand where evidence is obtained of distinct phylogenies among member species.

Virus Species

The species taxon has always been regarded as the most important hierarchical level in classification, but with the viruses it has proved to be the most difficult to deal with. After years of controversy, in 1991, the ICTV accepted the definition of a virus species proposed by van Regenmortel (1990), as follows: "A virus species is defined as a polythetic class of viruses that constitutes a replicating lineage and occupies a particular ecological niche." Members of a polythetic class are defined by more than one property and no single property is essential or necessary. One major advantage in this definition is that it can accommodate the inherent variability of viruses and it does not depend on the existence of a single unique characteristic. Similarly, it can accommodate the different traditions of virologists working in different areas of virology, in some cases accommodating "the lumpers" and in others "the splitters."

The ICTV Study Groups are now determining the specific properties to be used to define species in the taxon for which they are responsible. It seems clear that the species term will eventually be defined somewhat similarly to the term virus - although in some cases the term virus matches best with subspecies, strain or even variant. Just as the term virus is defined differently in different virus families, so, species will be defined differently, in some cases with emphasis on genome properties, and in others on structural, physicochemical or serological properties. Some viruses have already been designated as species, for example: Sindbis virus, Newcastle disease virus, poliovirus 1, vaccinia virus, Fiji disease virus, tomato spotted wilt virus. These examples, however, do not reflect the difficulty that is being encountered in deciding whether a particular virus should be designated as a species or as a subspecies or strain or variant.

Virus Nomenclature

Usage of Formal Taxonomic Nomenclature

In formal taxonomic usage, the first letters of virus family, subfamily, and genus names are capitalized and the terms are printed in italics (underlined when typewritten). Species designations are not capitalized (unless they are derived from a place name or a host family or genus name), nor are they italicized. In formal usage, the name of the taxon should precede the term for the taxonomic unit; for example:"the family Paramyxoviridae","the genus Morbillivirus." Furthermore, it was decided years ago that virus nomenclature would not involve the use of latinized binomial terms. For example, terms such as Flavivirus fabricis, Orthopoxvirus variolae and Herpesvirus varicellae, which were used at one time, have been abandoned. The following represent examples of full formal taxonomic terminology:

1. Family Poxviridae, subfamily Chordopoxvirinae, genus Orthopoxvirus, vaccinia virus.
2. Family Herpesviridae, subfamily Alphaherpesvirinae, genus Simplexvirus, human herpes virus 2 (herpes simplex virus 2).
3. Family Picornaviridae, genus Enterovirus, poliovirus 1.
4. Order Mononegavirales, Family Rhabdoviridae, genus Lyssavirus, rabies virus.
5. Family Bunyaviridae, genus Tospovirus, tomato spotted wilt virus.
6. Family Bromoviridae, genus Bromovirus, brome mosaic virus.
7. Genus Sobemovirus, Southern bean mosaic virus.
8. Family Totiviridae, genus Totivirus, Saccharomyces cerevisiae virus L-A.
9. Family Tectiviridae, genus Tectivirus, enterobacteria phage PRD1.
10. Family Plasmaviridae, genus Plasmavirus, Acholeplasma phage L2.

Vernacular Usage of Virus Nomenclature

In informal vernacular usage, virus family, subfamily, genus and species names are written in lower case Roman script; they are not capitalized, nor are they printed in italics or underlined. In informal usage, the name of the taxon should not include the formal suffix, and the name of the taxon should follow the term for the taxonomic unit; for example,"the picornavirus family", "the enterovirus genus." The use of vernacular terms for virus taxonomic units and virus names should not lead to unnecessary ambiguity or loss of precision in virus identification. The formal family, subfamily, and genus terms and standard ICTV vernacular species terms, rather than any synonyms or transliterations, should be used as the basis for choosing vernacular terms. One particular source of ambiguity in vernacular nomenclature lies in the common use of the same root terms in formal family and genus names. Imprecision stems from not being able to easily identify in vernacular usage which hierarchical level is being cited. For example, the vernacular name "paramyxovirus" might refer to the family Paramyxoviridae, the genus Paramyxovirus, or one of the species in the genus Paramyxovirus, such as one of the human parainfluenza viruses. Some virologists have suggested that this problem be solved by renaming taxa so that the same root term is never used at multiple hierarchical levels; however, there is no consensus for this, in fact, as plant virus taxonomy switches away from groups and toward families and genera, this problem will be exacerbated. The solution in vernacular usage is to avoid "jumping" hierarchical levels and to add taxon identification wherever needed. For example, when citing the taxonomic placement of human parainfluenza virus 1, the term "paramyxovirus" should refer firstly to the genus, not the subfamily or family, and taxon identification should always be added: "human parainfluenza virus 1 is a member of the paramyxovirus genus," rather than " human parainfluenza virus 1 is a paramyxovirus." Most examples like this exemplify the advantage of switching, where necessary, into formal nomenclature usage: "human parainfluenza virus 1 is a species in the genus Paramyxovirus, family Paramyxoviridae." In this example, as is usually the case, adding the information that this virus is also a member of the subfamily Paramyxovirinae and the order Mononegavirales is unnecessary.

Structural, Genomic, Physicochemical and Replicative Properties of Viruses Used in Taxonomy

The way, by which viruses are characterized, for taxonomic and other purposes, is changing rapidly. In the past, laboratory techniques have included characterizations of virion morphology (by electron microscopy), virion stability (by varying pH and temperature, adding lipid solvents and detergents, etc.), virion size (by filtration through fibrous and porous microfilters), and virion antigenicity (by many different serologic methods). These means worked because after large numbers of viruses had been studied and their characteristics placed into the universal taxonomic scheme, it was necessary in most cases to only measure a few characteristics to place a new virus, especially a new variant from a well studied source, in its proper taxonomic niche. For example, a new adenovirus, isolated from the human respiratory tract and identified by serologic means, was easy to place in its niche in the family Adenoviridae, genus Mastadenovirus. The exceptions occurred when a new virus was found that did not have a familiar set of properties. Such a virus became a candidate prototype for a new taxon, generally a new family or genus. In such cases, comprehensive characterization of all virion properties was called for.

One particularly important technological advance underpinning the development of modern virus taxonomy was the invention by Brenner and Horne (1959) of the negative staining technique for electron microscopic examination of virions. The impact of this technique was immediate:

(a) Virions could be characterized with respect to size, shape, surface structure, presence or absence of an envelope, and, often, symmetry.

(b) The method could be applied simply and universally.

(c) Virions could be characterized in unpurified material, including diagnostic specimens.

Negative staining has facilitated the rapid accumulation of data about the physical properties of many viruses. Thin-section electron microscopy of virus-infected cell cultures and tissues of infected humans, animals (including experimental animals), and plants has provided complementary data on virion morphology, mode and site of virion morphogenesis (e.g., site of budding), etc. Thus, in many cases, viruses were placed in their appropriate family, and in some instances in their appropriate genus, after simple visualization and measurement by negative-stain and/or thin-section electron microscopy.

The fundamental molecular bases for many of the empirical virion property measurements that were originally used to construct virus families and genera are now rather well understood. Many of the characteristics that have been used in deciding taxonomic constructions are followed.

Some Properties of Viruses Used in Taxonomy

Morphology

1. Virion size
2. Virion shape
3. Presence or absence and nature of peplomers
4. Presence or absence of an envelope
5. Capsid symmetry and structure

Physicochemical and Physical properties

1. Virion molecular mass
2. Virion buoyant density (in cscl, sucrose, etc.)
3. Virion sedimentation coefficient

4. Physical stability
5. Thermal stability
6. Cation stability (Mg++, Mn++)
7. Solvent stability
8. Detergent stability
9. Irradiation stability
10. Genome
11. Type of nucleic acid (DNA or RNA)
12. Size of genome in kb/kbp
13. Strandedness - (single) stranded or (double) stranded
14. Linear or circular
15. Sense (positive-sense, negative-sense, ambisense)
16. Number and size of segments
17. Nucleotide sequence, or partial sequence
18. Presence of repetitive sequence elements
19. Presence of isomerization
20. G+C ratio
21. Presence or absence and type of 5'-terminal cap
22. Presence or absence of 5'-terminal covalently-linked protein
23. Presence or absence of 3'-terminal poly (A) tract

Proteins

1. Number, size and functional activities of structural proteins
2. Number, size and functional activities of non-structural proteins
3. Details of special functional activities of proteins: especially transcriptase, reverse
4. Transcriptase, hemagglutinin, neuraminidase, and fusion activities
5. Amino acid sequence or partial sequence
6. Glycosylation, phosphorylation, myristylation
7. Epitope mapping

Lipids

1. Content, character, etc.

Carbohydrates

1. Content, character, etc.
2. Genome organization and replication

3. Genome organization
4. Strategy of replication
5. Number and position of open reading frames
6. Transcriptional characteristics
7. Translational characteristics
8. Post-translational processing
9. Site of accumulation of virion proteins
10. Site of virion assembly
11. Site and nature of virion maturation and release

Antigenic properties

1. Serologic relationships, especially as obtained inreference centers

Biologic properties

1. Natural host range
2. Mode of transmission in nature
3. Vector relationships
4. Geographic distribution
5. Pathogenicity, association with diseasetissue tropisms, pathology, histopathology

Through the use of monoclonal antibodies, synthesized peptides, and epitope mapping, there is new understanding of the molecular bases for those serological reactions that were originally used to construct families and genera. Today, genome sequencing, or partial sequencing, is often done very early in virus identification, and even in diagnostic activities. For comparison, genome sequences are available in readily accessible databases for the prototype viruses of nearly all taxa. Sequence data are even driving consideration for the construction of new families and new genera before other data are available. It is likely because of their absolute nature that genome sequence data will become the base for further refinement and expansion of the universal taxonomic scheme.

Table 3.1. The ssDNA Viruses

Family	*Subfamily*	*Genus*	*Type species*	*Host*
Inoviridae		Inovirus	Enterobacteria phage M13	Bacteria
		Plectrovirus	Acholeplasma phage MV-L51	Mycoplasma
Microviridae		Microvirus	Enterobacteria phage ØX174	Bacteria
		Spiromicrovirus	Spiroplasma phage 4	Spiroplasma
Geminiviridae		Mastrevirus	Maize streak virus	Plants
Circoviridae		Circovirus	Chicken anemia virus	Vertebrates
		Nanovirus	Subterranean clover stunt virus	Plants
Parvoviridae	Parvovirinae	Parvovirus	Mice minute virus	Vertebrates
		Erythrovirus	B19 virus	Vertebrates
		Dependovirus	Adeno-associated virus 2	Vertebrates
	Densovirinae	Densovirus	Junonia coenia densovirus	Invertebrates
		Iteravirus	Bombyx mori densovirus	Invertebrates
		Brevidensovirus	Aedes aegypti densovirus	Invertebrates

Table 3.2. The Negative Stranded ssRNA Viruses (Order Mononegavirales)

Family	*Subfamily*	*Genus*	*Type species*	*Host*
Bornaviridae		Bornavirus	Borna disease virus	Vertebrates
Filoviridae		Ebola-like viruses	Ebola virus	Vertebrates
		Marburg-like viruses	Marburg virus	Vertebrates
Paramyxoviridae	Paramyxovirinae	Respirovirus	Human parainfluenza virus I	Vertebrates
		Morbillivirus	Measles virus	Vertebrates
		Rubulavirus	Mumps virus	Vertebrates
	Pneumovirinae	Pneumovirus	Human respiratory syncytial virus	Vertebrates
		Metapneumovirus	Turkey rhinotracheitis virus	Vertebrates
Rhabdoviridae		Vesiculovirus	Vesicular stomatitis Indiana virus	Vertebrates
		Lyssavirus	Rabies virus	Vertebrates
		Ephemerovirus	Bovine ephemeral fever virus	Vertebrates
		Novirhabdovirus	Infectious hematopoietic necrosis virus	Vertebrates

Continue...

	Cytorhabdovirus	Lettuce necrotic yellows virus	Plants
	Nucleorhabdovirus	Potato yellow dwarf virus	Plants
Orthomyxoviridae	Influenzavirus A	Influenza A virus	Vertebrates
	Influenzavirus B	Influenza B virus	Vertebrates
	Influenzavirus C	Influenza C virus	Vertebrates
	Thogotovirus	Thogoto virus	Vertebrates
Bunyaviridae	Bunyavirus	Bunyamwera virus	Vertebrates
	Hantavirus	Hantaan virus	Vertebrates
	Nairovirus	Nairobi sheep disease virus	Vertebrates
	Phlebovirus	Sandfly fever Sicilian virus	Vertebrates
	Tospovirus	Tomato spotted wilt virus	Plants
	Tenuivirus	Rice stripe virus	Plants
	Ophiovirus	Citrus psorosis virus	Plants
Arenaviridae	Arenavirus	Lymphocytic choriomeningitis virus	Vertebrates
	Deltavirus	Hepatitis delta virus	Vertebrates

Table 3.3. The Positive Stranded ssRNA Viruses:

Family	*Genus*	*Type species*	*Host*
Narnaviridae	Narnavirus	Saccharomyces cerevisiae 20S narnavirus	Yeast
	Mitovirus	Cryphonectria parasitica NB631 virus	Yeast
Leviviridae	Levivirus	Enterobacteria phage MS2	Bacteria
	Allolevivirus	Enterobacteria phage Qb	Bacteria
Picornaviridae	Enterovirus	Poliovirus 1	Vertebrates
	Rhinovirus	Human rhinovirus 1A	Vertebrates
	Hepatovirus	Hepatitis A virus	Vertebrates
	Cardiovirus	Encephalomyocarditis virus	Vertebrates
	Aphthovirus	Foot-and-mouth disease virus O	Vertebrates
	Parechovirus	Human echovirus 22	Vertebrates
	Cricket paralysis-like viruses	Cricket paralysis virus	Invertebrates
Sequiviridae	Sequivirus	Parsnip yellow fleck virus	Plants
	Waïkavirus	Rice tungro spherical virus	Plants
Comoviridae	Comovirus	Cowpea mosaic virus	Plants
	Fabavirus	Broad bean wilt virus 1	Plants

Continue...

	Nepovirus	Tobacco ringspot virus	Plants
Potyviridae	Potyvirus	Potato virus Y	Plants
	Rymovirus	Ryegrass mosaic virus	Plants
	Macluravirus	Maclura mosaic virus	Plants
	Ipomovirus	Sweet potato mild mottle virus	Plants
	Bymovirus	Barley yellow mosaic virus	Plants
	Tritimovirus	Wheat streak mosaic virus	Plants
Caliciviridae	Vesiculovirus	Swine vesicular exanthema virus	Vertebrates
	Lagovirus	Rabbit hemorrhagic disease virus	Vertebrates
	Norwalk-like viruses	Norwalk virus	Vertebrates
	Sapporo-like viruses	Sapporo virus	Vertebrates
	Hepatitis E -like viruses	Hepatitis E virus	Vertebrates

Table 3.4. The Subviral Agents: Satellites, Viroids, and Agents of Spongiform Encephalopathies (Prions)

The Subviral Agents	*Family*	*Genus*	*Type species*	*Host*
Satellites			Tobacco necrosis virus satellite	Plants Vertebrates Invertebrates Fungi
Viroids	Pospiviroidae	Pospiviroid	Potato spindle tuber viroid	Plants
		Hostuviroid	Hop stunt viroid	Plants
		Cocadviroid	Coconut cadang-cadang viroid	Plants
		Apscaviroid	Apple scar skin viroid	Plants
		Coleviroid	Coleus blumei viroid 1	Plants
	Avsunviroidae	Avsunviroid	Avocado sunblotch viroid	Plants
		Pelamoviroid	Chrysanthemum chlorotic mottle viroid	Plants
Prions			Scrapie agent Vertebrates	Fungi

4

PLANT VIRUSES

In 1892, the Russian virologist D. Iwanowski discovered horizontal transmission of viruses. He succeeded in infecting plants with the tobacco mosaic disease by applying a bacteria-free sap gained from infected tobacco leaves. The Dutchman M.W. Beijerinck confirmed this discovery some six years later. He filtered an extract of infected plants through a porcelain strainer not viable for bacteria, and found that the sap thus obtained displayed no loss of its infecting ability. The virus itself was not isolated till 1935 and not crystallized before 1937. In 1937, F.C. Bawden and N.W. Pirie (Rothamsted Experimental Station, England) detected that the preparations contained phosphate, an inherent part of the RNA molecule.

A number of TMV wild types differing, for example, in their host range or in the primary structure of their coat proteins exist. The classic strain that at the same time is the one that proliferates best on tobacco plants is called vulgare. A strain from tomatoes that proliferates equally well on tobacco plants is called dahlemense A third strain gained from plantain is known as Holmes ribgrass. The single strains differ visibly in the symptoms they cause in tobacco plants. The amino acid sequences of the coat proteins of four different wild strains show that a certain sequence remains always unchanged. This sequence is assumed to be in direct contact with the RNA. The analysis of the coats' tertiary structures confirmed this assumption.

Characteristics of Plant Viruses

Detection: Small-require electron microscope for visualization. ELISA (Enzyme Linked Immuno Sorbent Assay) is often used to detect.

Morphology

1. *Rods*: Rigid rods - 5 × 300 nm Flexuous rods - 10-13 × 480-2000 nm
2. *Spheres*: Icosadodecahedron - 60 nm. Satellite - 17 nm Membrane encased - 100 nm, Bacillus-like, cylindrical rods 52-75 × 300-800 nm
3. *Monopartite*: All particles of the virus are the same size and contain the same nucleic acid.
4. *Multipartite*: Typically di- or tri- partite - virus population is made of two or three, respectively, sizes of particle with distinct nucleic acid strands encased in each capsid.

Composition and Structure

Each plant virus consists of at least one nucleic acid and one protein. Some viruses also have enzymes, lipids, and carbohydrates. The nucleic acid makes up 5 to 40 percent of the virus with the remaining 95 to 60 percent being protein. The proportion of nucleic acid to protein is greatest in spherical viruses. This simplest virus has a genome of approximately 1 to 3 $\times 10^6$ daltons. This is in comparison with bacteria that have 1.5 $\times 10^9$ daltons.

Composition and Structure of Viral Protein

The proteins of plant viruses are made of the same amino acids common to plant proteins. There are no unique amino acids distinctive of plant viruses.

The protein shells of plant viruses are called capsids and are composed of repeating subunits. For any given virus the capsid proteins are identical and interchangeable though there is no universality of proteins between two different viruses. In many cases a single protein composes the entire capsid, though more complex viral morphologies may require multiple protein types for their capsids. An individual virion is composed of a specific number of proteins unique to that virus particle. This is true irrespective of the morphological type of the virion. Plant viruses encode specific proteins that are involved in a variety of transport processes such as intra- and intercellular transport of viral genomes, long distance movement and transmission of virus particles between plants by vectors such as insects. It has extensive experience in studying the molecular biology of luteoviruses (Potato leafroll virus; PLRV) and umbraviruses (Groundnut rosette virus, Pea enation mosaic virus -2) that allows to use these viruses as experimental systems for analysing functions and biochemical properties of different "transport" proteins. Cell-to-cell and long-distance movement proteins encoded by umbraviruses have been identified. The methods used to explore how these proteins operate in plants include mutagenesis, immunocytochemical electron microscopy to investigate intracellular location, including nuclear and nucleolar targeting, biochemical studies such as of RNA-binding including the visualisation of RNA-protein complexes using atomic force microscopy. These approaches have the promise of providing powerful ways to investigate the modulation of plant transport processes.

Composition and Structure of Viral Nucleic Acid

The majority of plant viruses contain their genomes encoded as single strand RNA (ssRNA). However, double strand RNA (dsRNA), single strand DNA (ssDNA), and double strand DNA (dsDNA) plant viruses also occur. The nucleotide bases that make up plant virus RNA and DNA are the same as those typical to RNA and DNA of their host plants.

Satellite Viruses and Satellite RNAs

Some viruses require helper viruses to infect plants. These helper viruses are termed Satellite Viruses. "Satellite viruses are not related, or are only partially related, to the RNA of the virus; satellite RNAs may increase or decrease the severity of viral infections."

Amplicon-mediated RNA Silencing

An amplicon is a transgene that comprises a viral genome. The first to be produced was that of Tobacco mosaic virus (Yamaya et al., Mol. Gen. Genetics, 1988, 211, 520-525). Subsequently, using Potato virus X, Angell and Baulcombe (EMBO Journal, 1997, 16, 3675-3684) described how amplicons were capable of inducing RNA silencing.

Amplicons that consist of a full-length cDNA of the Potato leafroll virus (PLRV) genome or the same cDNA modified to contain DNA encoding green fluorescent protein (PLRV-GFP). PLRV and PLRV-GFP transgene-derived transcripts can infect protoplasts and thus because every cell of a transformed plant produces transcript, each should generate infectious RNA. However, replication of the transcribed amplicon was found to induce strong RNA silencing, as evidenced by the absence of virus symptoms, accumulation of only small amounts of PLRV-GFP RNA, low and non-uniform PLRV distribution and GFP expression, and accumulation of small 21-25 nt RNAs. These amplicons are being used to study the regulation of virus gene expression and the mechanisms of RNA silencing. This work is revealing the strategies used by viruses to combat host defence based on RNA silencing.

Biological Function of Viral Components

Coding

The function of the protein coat is not limited to encasing the viral genome. The outer surface of the protein is a major determinant in vector transmissibility. Barley Stripe Mosaic Virus contains a small carbohydrate component as a glycoprotein on the surface of its rod shaped capsid. This surface carbohydrate has been shown to be bound by lectins in barley seed and is thought to play a role in seed transmission of this virus.

Infectivity is strictly the province of the nucleic acid. In fact, under experimental conditions the nucleic acid, extracted away from its capsid, may be used to inoculate plants. Redundancy is not a feature of the viral genome. The amount on information necessary to code for the capsid protein is small (474 nucleotides for the 158 amino acids of Tobacco Mosaic Virus) relative to the total viral genome (6400 nucleotides); therefore it is obvious that the genome contains more information that that required to produce nucleocapsid proteins. The TMV genome codes for; nucleocapsid protein, two viral replicase (make more viral genome) enzymes, and a protein that facilitates movement through the plasmodesmata.

There is no direct evidence for coding for disease causing factors and it has been assumed that symptoms that are produced the result of disrupting host cellular process rather than specific viral metabolites.

There are three major groups of plant viruses that seem to be most important among new disease-causing agents worldwide and that could reasonably be termed "emerging" in terms of apparently being new viruses causing new and serious diseases. The most important of these are in the taxonomic families Potyviridae and Geminiviridae, the Tospovirus genus of the Bunyaviridae, and perhaps also pararetroviruses like Banana streak virus (BSV).

Potyviridae

The Potyviridae are the largest single taxonomic group of plant viruses, and the subject of more scientific papers per year than any other plant virus taxonomic group except perhaps the geminiviruses. It seems that potyviruses, and specifically aphid-transmitted viruses of the genus *Potyvirus*, are one of the most successful groups of plant pathogens in the world. The genus has been claimed to be almost as ancient as flowering plants, and has a worldwide distribution throughout higher plants. New potyviruses seem to appear with new crop introductions just about anywhere these occur. In South Africa, for example, *Passiflora edulis* (passionfruit) cultivation in the 1980s was severely limited by diseases, many of which appeared linked to a mixture of potyvirus-like agents. One of these viruses was characterized here in detail and found to be not the passionfruit woodiness virus (PWV) characterized in Australia and the Far East, but a distinct species now known to be a strain of cowpea aphid-borne mosaic virus (CABMV). It is very interesting that a similar disease in Brazil seems to be caused by a closely-related strain of the same virus, whereas woodiness disease in Australasia and Asia is caused by distinct species of viruses.

Another intriguing aspect of emerging potyviral plant diseases is the appearance of hitherto latent virus infections. A well-documented case investigated here was the appearance in cultivated specimens of the horticulturally important flowering bulbs Ornithogalum and Lachenalia species of at least two distinct potyviruses. One of these, Ornithogalum mosaic virus (OrMV), first described in South African-derived bulbs in Holland, was found in plant specimens collected in localities where the plants were endemic. However, it caused apparent or overt infections only when plants were kept in cultivation.

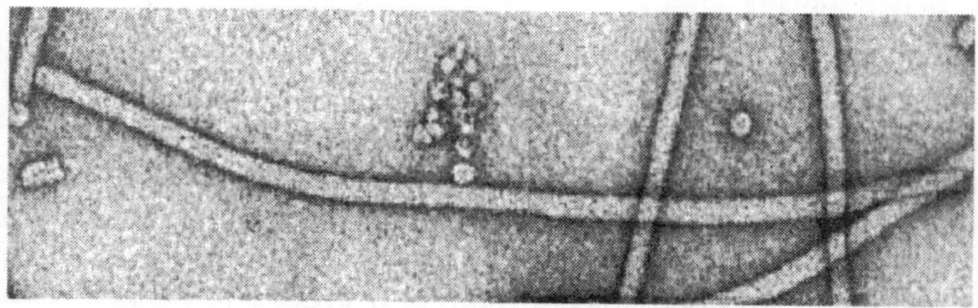

Fig. 4.1. Particle of ornithogalum mosaic virus (OrMV)

Geminiviridae

Viruses in the family Geminiviridae, and specifically the whitefly-transmitted begomoviruses, are presently associated with severe diseases in tomatoes the world over, with cotton disease in India and elsewhere, and with disease problems and food shortages due to disease in cassava in central Africa. Infections with Bean golden mosaic virus (BGMV) are apparently the single largest limiting factor to bean production in Central America

Polston and Anderson (1997) covered the Caribbean basin and the Central American region in an excellent review of the emergence of whitefly-transmitted geminiviruses (WTGs) in tomatoes. They estimated that 20 - 100% of crop loss throughout these areas has been caused by epidemics of WTGs, ranging from Tomato golden mosaic (TGMV) to Tomato yellow leaf curl (TYLCV) to Potato yellow mosaic (PYMV). They listed 17 distinct viruses known to affect tomatoes in this region, and this is probably a conservative estimate, as more are constantly being found.

Much of the apparently emerging nature of the virus diseases is due to the worldwide spread of a new "silverleaf" or B biotype of the vector *Bemisia tabaci*, which is now being touted as the new species Bemisia argentifolii. The new vector has a much wider range of preferences for feeding than older vector types, which has apparently resulted in the spread of viruses that normally infect only weed or endemic plant species into adjacent, previously untargeted crop species. Roye et al. (1997) surveyed geminiviruses in weeds (notably *Sida* and *Wissadula* species) in Jamaica and showed a population of viruses distinct from crop-infecting geminiviruses - and one perhaps poised to enter crops once a suitable vector biotype appears. Although the problem also exists in the developed world , and most notably in the southern United States and Europe, it can be dealt with, to at least some extent, by changing cultivation practices. The same option does not exist in the developing world, however, due to the expensive nature of the measures (screen houses, heavy spraying regimes). Thus, the inexorable spread of the vector into these areas will almost certainly mean heavy crop losses.

Begomovirus diseases also seem to exemplify a phenomenon first noted with potyviruses; unrelated or distinct virus species in the genus cause very similar disease syndromes in the same crop plants, usually in widely separated geographical areas but sometimes in the same small area. Thus, TYLCV-Israel, -Sardinia, and -Thailand are distinct species of virus despite causing the same disease; however, so are the several viruses causing cotton leaf curl disease in India and Pakistan, and the several distinct viruses causing tomato leaf curl in India. This is undoubtedly related to their worldwide distribution and also to their apparent antiquity. Geminivirus variation appears linked to geographic separation and host plant genetic divergence, as well as to continental drift (Rybicki, 1994). All these factors indicate that the youngest age for the family Geminiviridae is hundreds of millions of years and that of begomoviruses is at least tens of millions of years. This diversity and ubiquity guarantee that a wide variety of virus genotypes exists, probably as well-adapted and essentially symptomless infections in endemic species or weeds, and that any change in transmission characteristics of the vector will result in appearance of the virus in crop plants almost instantly.

Whereas begomoviruses have been making their presence felt as they spread out of endemic hosts into cultivated species, viruses in the genus *Mastrevirus* continue to emerge into crop species in Africa. Maize streak disease (MSD) was first described in 1901, so can hardly be considered as emerging. However, more recent findings indicate that a hitherto unsuspected and distinct group of Maize streak virus (MSV) strains cause disease in wheat and some grasses, rather than the disease being caused by the same closely-related group of virus isolates and strains as are found in maize (Rybicki et al., 1998; see also here). Sugarcane streak virus(es) (SSVs) also still seem to be present in Africa. Whereas the better characterized variants SSV-Natal and SSV-Mauritius have not been a problem in southern Africa for over 30 years, and especially not in commercially-grown sugarcane varieties, distinct mastrevirus species have recently been characterized from mildly diseased "traditional" material from Egypt (Bigarré et al., 1999).

A ray of hope in the struggle to combat TYLCV-like diseases is the fact that genetic engineering approaches using pathogen-derived resistance appear to be quite promising. Tomato plants transgenic for the TYLCV coat protein appear to be resistant to the virus (Kunik et al., 1994). Additionally, transgenic Nicotiana benthamiana plants expressing antisense RNA to the Rep gene were resistant to TYLCV infection (Bendahmane and Gronenborn, 1997). These developments, coupled with traditional and marker-assisted breeding techniques, may allow rapid dissemination of improved material to farmers.

Tospoviruses

The genus Tospovirus (family Bunyaviridae) is an extremely important group of plant viruses capable of infecting a large range of important crops. The type member, Tomato spotted wilt virus (TSWV), has one of the broadest host ranges among plant viruses. Tospoviruses are transmitted exclusively by thrips in a circulative propagative manner, meaning that the virus multiplies in the vector. TSWV was thought to be the only member of this genus until recently, and has had a serious impact on many crop species worldwide. It causes severe outbreaks in a large variety of crops grown in tropical and subtropical climate zones. Since 1990, other related but distinct viruses in this genus have been identified. The following members or tentative members are described: TSWV, Groundnut ringspot virus (GRSV), Tomato chlorotic spot virus (TCSV), Impatiens necrotic spot virus (INSV), Watermelon silver mottle virus (WSMV), and Groundnut bud necrosis virus (GBNV).

Owing to the worldwide spread of the Western flower thrip (Frankliniella occidentalis Pergande), the most efficient vector of tospoviruses, TSWV and some of the other tospoviruses are increasing being reported as causing problems in developing countries. These "new" tospoviruses often originate in developing countries, and some appear to occur only in these countries. For example, GRSV was first detected on peanuts during a survey of the viruses of this crop in South Africa, and isolate SA-05 now serves as the type isolate of GRSV. Thus far, this virus has been detected only on peanuts in South Africa, where it is relatively common, and on tomatoes in Brazil and Argentina. TCSV was first reported from Brazil, and GBNV was first detected in India.

Research on this important virus group has increased dramatically, and new tospoviruses are constantly being found. It could be that only the lack of resources in developing countries for the diagnosis of hard-to-identify viruses prevents even more reports of new tospoviruses.

Banana Streak Virus/Pararetroviruses

Banana streak virus (BSV) is a pararetrovirus - that is, a virus with circular dsDNA in virions which replicates via a longer-than-genome-length ssRNA(+) intermediate, via reverse transcription (see here). BSV may be considered an emerging problem in light of recent findings suggesting, first, that it occurs in an integrated form in Musa (see here); second, that it has been found in this form in all Musa germplasm tested thus far; third, that it is seed transmitted; and fourth, that the integrated virus can be activated to form the episomal form through stress. Stress can possibly include tissue culture propagation. Thus, it may be that initial BSV infections may largely be the result of plant stress, which means that BSV infections could occur anywhere, at any time, in the absence of any vectored introduction. It may be that genetic engineering in the form of a "knock-out" of the integrated viral genome can come to the rescue, at least of major commercial varieties.

Virus Infection of Plants

In economic terms, viruses are only of importance if it is likely that they will spread to crops during their commercial lifetime, which of course varies greatly between very short extremes in horticultural production and very long extremes in forestry. Some estimates put total worldwide damage due to plant viruses as high as US\$ 6×10^{10} per year.

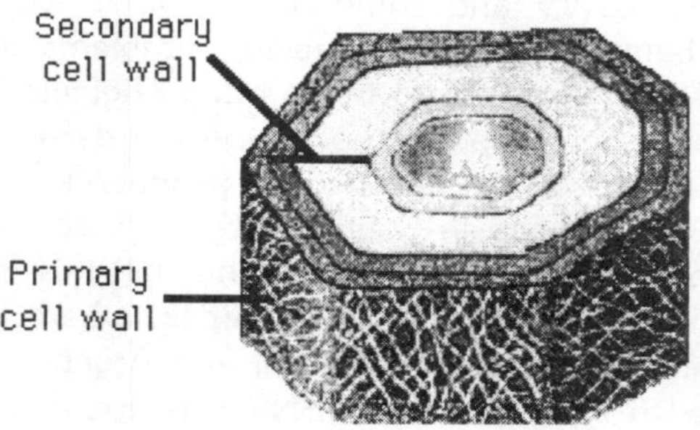

Fig. 4.2. T.S. of stem.

Plant viruses face special problems initiating an infection. The outer surfaces of plants are composed of protective layers of waxes and pectin, but more significantly, a thick wall of cellulose overlying the cytoplasmic membrane surrounds each cell.

To date, no plant virus is known to use a specific cellular receptor of the type that animal and bacterial viruses use to attach to cells. Rather, plant viruses rely on a mechanical breach of the integrity of a cell wall to directly introduce a virus particle into a cell. This is achieved either by the vector associated with transmission of the virus or simply by mechanical damage to cells. After replication in an initial cell, the lack of receptors poses special problems for plant viruses in recruiting new cells to the infection.

Transmission of Plant Viruses

There are a number of routes by which plant viruses may be transmitted:

Seeds

These may transmit virus infection either due to external contamination of the seed with virus particles, or due to infection of the living tissues of the embryo. Transmission by this route leads to early outbreaks of disease in new crops, which are usually initially focal in distribution, but may subsequently be transmitted to the remainder of the crop by other mechanisms.

Vegetative propagation/grafting

These techniques are cheap and easy methods of plant propagation but provide the ideal opportunity for viruses to spread to new plants.

Vectors

Many different groups of living organisms can act as vectors and spread viruses from one plant to another:

- Bacteria (e.g. Agrobacterium tumefaciens - the Ti plasmid of this organism has been used experimentally to transmit virus genomes between plants)
- Fungi
- Nematodes
- Arthropods: Insects - aphids, leafhoppers, planthoppers, beetles, thrips, etc.
- Arachnids - mites

Fig. 4.3. Vectors.

Mechanical

Mechanical transmission of viruses is the most widely used method for experimental infection of plants and is usually achieved by rubbing virus-containing preparations into the leaves, which in most plant species are particularly susceptible to infection. However, this is also an important natural method of transmission. Virus particles may contaminate soil for long periods and may be transmitted to the leaves of new host plants as wind-blown dust or as rain-splashed mud.

> Transmission of plant viruses by insects is of particular agricultural importance. Extensive areas of monoculture and the inappropriate use of pesticides which kill natural predators can result in massive population booms of insects such as aphids.
>
> Plant viruses rely on a mechanical breach of the integrity of a cell wall to directly introduce a virus particle into a cell. This is achieved either by the vector associated with transmission of the virus or simply by mechanical damage to cells.

Transfer by insect vectors is a particularly efficient means of virus transmission. In some instances, viruses are transmitted mechanically from one plant to the next by the vector and the insect is merely a means of distribution, flying or being carried on the wind for long distances (sometimes hundreds of miles). Insects which bite or suck plant tissues are, of course, the ideal means of transmitting viruses to new hosts. This is known as non-propagative transmission. However, in other cases (e.g. many plant rhabdoviruses) the virus may also infect and multiply in the tissues of the insect (propagative transmission) as well as those of host plants. In these cases, the vector serves as a means not only of distributing the virus, but also of amplifying the infection:

Group III geminivirus are transmitted by insect vectors (leafhoppers or whiteflies) and their genomes consist of two circular, single-stranded DNA molecules. These viruses cause a great deal of crop damage in plants such as tomatoes, beans, squash, cassava and cotton and their spread may be directly linked to the inadvertent world-wide dissemination of a particular biotype of the whitefly *Bemisia tabaci* . This vector is an indiscriminate feeder, encouraging rapid and efficient spread of viruses from indigenous plant species to neighbouring crops.

Multipartite Plant Viruses

- Segmented virus genomes are those which are divided into two or more physically separate molecules of nucleic acid, all of which are then packaged into a single virus particle.

- Although multipartite genomes also segmented, each genome segment is packaged into a separate virus particle.

Family	Segments
Geminivirus (group III) (single-stranded DNA)	Bipartite
Comovirus (single-stranded RNA)	Bipartite
Furovirus (single-stranded RNA)	Bipartite
Tobravirus (single-stranded RNA)	Bipartite
Partitiviridae (double-stranded RNA)	Bipartite
Bromoviridae (single-stranded RNA)	Tripartite
Hordeivirus (single-stranded RNA)	Tripartite

These discrete particles are structurally similar and may contain the same component proteins, but often differ in size depending on the length of the genome segment packaged. In one sense, multipartite genomes are, of course, segmented, but this is not the strict meaning of these terms as they are used here.

Genome segmentation reduces the probability of breakages due to shearing, thus increasing the total potential coding capacity of the genome. However, the disadvantage of this strategy is that all the individual genome segments must be packaged into each virus particle, or the virus will be defective as a result of loss of genetic information. Separating the genome segments into different particles (the multipartite strategy) removes the requirement for accurate sorting, but introduces a new problem in that all the discrete virus particles must be taken up by a single host cell to establish a productive infection. This is perhaps the reason multipartite viruses are only found in plants. Many of the sources of infection by plant viruses, such as inoculation by sap-sucking insects or after physical damage to tissues, result in a large inoculum of infectious virus particles, providing opportunities for infection of an initial cell by more than one particle.

Pathogenesis of Plant Virus Infections

Initially, most plant viruses multiply at the site of infection, giving rise to localized symptoms such as necrotic spots on the leaves. Subsequently, the virus may be distributed to all parts of the plant either by direct cell-to-cell spread or by the vascular system, resulting in a systemic infection involving the whole plant. However, the problem these viruses face in reinfection and recruitment of new cells is the same as they face initially - how to cross the barrier of the plant cell wall.

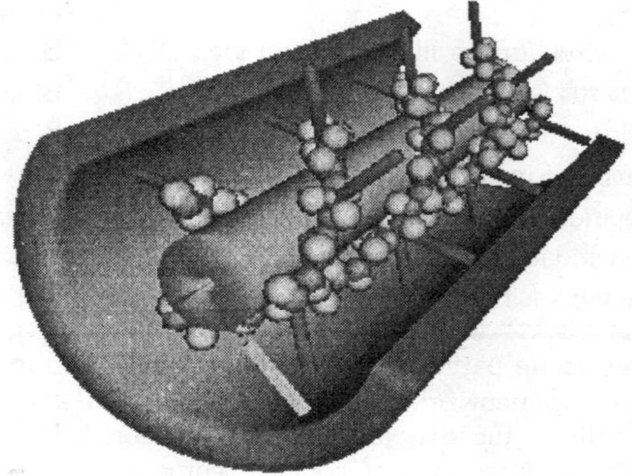

Fig. 4.4. Virus within plant cell.

Plant cell walls necessarily contain channels called plasmodesmata which allow plant cells to communicate with each other and to pass metabolites between them. However, these channels are too small to allow the passage of virus particles or genomic nucleic acids.

Many (if not most) plant viruses have evolved specialized movement proteins which modify the plasmodesmata. One of the best known examples of this is the 30k protein of tobacco mosaic virus (TMV). This protein is expressed from a sub-genomic mRNA and its function is to modify plasmodesmata causing genomic RNA coated with 30k protein to be transported from the infected cell to neighbouring cells. Other viruses, such as cowpea mosaic virus (CPMV - Comovirus family) have a similar strategy but employ a different molecular mechanism. In CPMV, the 58/48k proteins form tubular structures allowing the passage of intact virus particles to pass from one cell to another.

Typically, virus infections of plants might result in effects such as growth retardation, distortion, mosaic patterning on the leaves, yellowing, wilting, etc. These macroscopic symptoms result from:

Plants might be seen as sitting targets for virus infection - unlike animals, they cannot run away. However, plants exhibit a range of responses to virus infections designed to minimize their

Fig. 4.5. Viral symptoms in plant.

effects. Initially, infection results in a 'hypersensitive response', manifested as the synthesis of a range of new proteins, the 'PR' (pathogenesis related) protein'. Although this system is poorly understood, at least some of these proteins have been characterized and have been shown to be proteases, which presumably destroy virus proteins, limiting the spread of the infection. There is some similarity here between this response and the production of interferons by animals. In addition, systemic resistance to virus infection is a naturally occurring phenomenon in some strains of plant, e.g. the tobacco N gene encodes a cytoplasmic protein with a nucleotide-binding site, which interferes with the TMV replicase. This is clearly a highly desirable characteristic and is highly prized by plant breeders, who try to spread this attribute to economically valuable crop strains.

Necrosis of cells, caused by direct damage due to virus replication

Hypoplasia, i.e. localized retarded growthfrequently leading to mosaicism (the appearance of thinner, yellow areas on the leaves)

Hyperplasia, which is excessive cell division or the growth of abnormally large cells, resulting in the production of swollen or distorted areas of the plant.

There are probably many different mechanisms involved in systemic resistance, but in general terms there is a tendency towards increased local necrosis as substances such as proteases and peroxidases are produced by the plant to destroy the virus and to prevent its spread and subsequent systemic disease. An example of this is the N gene, which when present in plants causes TMV to produce a localized, necrotic infection rather than the systemic mosaic symptoms normally seen.

Virus-resistant plants have been created by the production of transgenic plants expressing recombinant virus proteins or nucleic acids, which interfere with virus replication without producing the pathogenic consequences of infection, e.g:

1. Virus coat proteins: these have a variety of complex effects, including:
   - Inhibition of virus uncoating
   - Interference of expression of the virus at the level of RNA ("gene silencing" by 'untranslatable' RNAs)
2. Intact or partial virus replicases which interfere with genome replication
3. Antisense RNAs
4. Defective viral genomes
5. Satellite sequences
6. Catalytic RNA sequences (ribozymes)
7. Modified movement proteins

This is a very promising technology which offers the possibility of substantial increases in agricultural production without the use of expensive, toxic and ecologically damaging chemicals (fertilizers, herbicides, or pesticides), but is at present still in its infancy.

Another developing aspect of plant biotechnology is the construction of CVPs - Chimaeric Virus Particles - recompinant viruses such as:

- Cowpea mosaic virus - an icosahedral member of the Comovirus family

  or

- Potato X virus- a rod-shaped member of the Potexvirus family which express short antigenic peptides on their surface. Several grams of CVPs can be produced cheaply and easily per kilogram of plant leaves. This is a very promising vaccine technology.

Emergent Plant Viruses

Rarely, there seems to be an example of an emergent virus which has acquired extra genes and as a result of new genetic capacity, becomes capable of infecting new species. A possible example of this phenomenon is seen in tomato spotted wilt virus (TSWV). TSWV is a Bunyavirus with a very wide plant host range, infecting over 600 different species from 70 families. In recent decades, this virus has been a major agricultural pest in Asia, the Americas, Europe Africa. Its rapid spread has been due to dissemination of its insect vector (the thrip Frankinellia occidentalis) and diseased plant material. TSWV is the type species of the Tospovirus genus and has a similar morphology and genomic organization to the other Bunyaviruses. However, TSWV undergoes propagative transmission and it has been suggested that it may have acquired an extra gene in the M segment via recombination, either from a plant or from another plant virus. This new gene encodes a movement protein, conferring the capacity to infect plants and cause extensive damage.

ICTV Approved Plant Virus Families and Genera

This list is a rearranged list of a large database containing descriptions of all viruses at the Australian National University's Bioinformatics Facility developed and maintained at the Research School of Biological Sciences. The rearangement divided plant viruses into serveral categories according to the type of viral genomic nucleic acids. The original list of the taxonomic structure of virus families and genera has been compiled from:

The Classification and Nomenclature of Viruses "Sixth Report of the International ommittee on Taxonomy of Viruses" (1995) Eds. F.A. Murphy, C.M. Fauquet, M.A. Mayo, A.W. Jarvis, S.A. Ghabrial, M.D. Summers, G.P. Martelli, D.H.L. Bishop Archives of Virology, Springer Verlag, Wien New York

Table 4.1. DNA Viruses

(A) Circular dsDNA Viruses

Family	*Genus*	*Type Species*
Caulimoviridae	Badnavirus	Commelina yellow mottle virus
	Caulimovirus	Cauliflower mosaic virus
	SbCMV-like viruses	Soybean chloroticmottle
	CsVMV-like viruses	virus
	RTBV-like viruses	Cassava vein mosaicvirus
	Petunia vein	Rice tungro bacilliformvirus
	clearing-like viruses	Petunia vein clearing virus

(B) Circular ssDNA Viruses

Family	*Genus*	*Type Species*
Geminiviridae	Mastrevirus	Maize streak virus
	(Subgroup I Geminivirus)	Beet curly top
	Curtovirus	Bean golden mosaic
	(Subgroup II Geminivirus)	virus
	Begomovirus	
	(Subgroup III Geminivirus)	

Throughout this document, each type species name is preceded by the name of the order, family and genus it belongs to. Names of orders, families and genera are in italic script, if they are approved by the ICTV. Taxa names in quotes (and not in italic s cript) indicate that this taxon has not an ICTV international approved name. Species (vernacular) names are given in regular script. Viruses with no formal assignment to genus or family are indicated.

Table 4.2. RNA Viruses

(A) ssRNA Viruses

Family	*Genus*	*Type Species*
Bromoviridae	Alfamovirus	Alfalfa mosaic virus
	Ilarvirus	Tobacco streak virus
	Bromovirus	Brome mosaic virus
	Cucumovirus	Cucumber mosaic virus
Closteroviridae	Closterovirus	Beet yellow virus
	Crinivirus	Lettuce infectious yellows virus
Comoviridae	Comovirus	Cowpea mosaic virus
	Fabavirus	Broad bean wilt virus
	Nepovirus	Tobacco ringspot virus
Potyviridae	Potyvirus	Potato virus Y
	Rymovirus	Ryegrass mosaic virus
	Bymovirus	Barley yellow mosaic virus
Sequiviridae	Sequivirus	Parsnip yellow fleck virus
	Waikavirus	Rice tungro spherical virus
Tombusviridae	Carmovirus	Carnation mottle virus
	Dianthovirus	Carnation ringspot virus
	Machlomovirus	Maize chlorotic mottle virus

	Necrovirus	Tobacco necrosis virus
	Tombusvirus	Tomato bushy-stunt virus
Unassigned Genera of ssRNA viruses	Capillovirus	Apple stem grooving virus
	Carlavirus	Carnation latent virus
	Enamovirus	Pea enation mosaic virus
	Furovirus	Soil-borne wheat mosaic virus
	Hordeivirus	Barley stripe mosaic virus
	Idaeovirus	Raspberry bushydwarf virus
	Luteovirus	Barley yellow dwarf virus
	Marafivirus	Maize ravado fino virus
	Potexvirus	Potato virus X
	Sobemovirus	Southern bean mosaic virus
	Tenuivirus	Rice stripe virus
	Tobamovirus	Tobacco mosaic virus
	Tobravirus	Tobacco rattle virus
	Trichovirus	Apple chlorotic leaf spot virus
	Tymovirus	Turnip yellow mosaic virus
	Umbravirus	Carrot mottle virus

(B) Negative ssRNA Viruses: Order - Mononegavirales

Family	*Genus*	*Type Species*
Rhabdoviridae	Cytorhabdovirus	Lettuce necrotic yellow virus
	Nucleorhabdovirus	Potato yellow dwarf virus
Bunyaviridae	Tospovirus	Tomato spotted wilt virus

(C) dsRNA Viruses:

Family	*Genus*	*Type Species*
Partitiviridae	Alphacryptovirus	White clover cryptic virus 1
	Betacryptovirus	White clover cryptic virus 2
Reoviridae	Fijivirus	Fiji disease virus
	Phytoreovirus	Wound tumor virus
	Oryzavirus	Rice ragged stunt virus

The Use of Phylogenetic Analysis in the Classification of Plant Viruses

The classification of plant viruses has traditionally been a conservative field of endeavour, with plant virologists resolutely resisting the introduction of Latinate nomenclature and Linnaean classification schemes. Plant viruses have also until recently been classified in only two taxa - the "virus" (more or less equivalent ot animal virus species) and the "group" (equivalent to genus).

Table 4.3. Unassigned Viruses.

Genome	*Species*
ssDNA	Banana bunchy top virus
	Coconut foliar decay virus
	Subterranean clover stunt virus
ds DNA	Cucumber vein yellowing virus
ds RNA	Tobacco stunt virus
ss RNA	Garlic viruses A,B,C,D
	Grapevine fleck virus
	Maize white line mosaic virus
	Olive latent virus 2
	Ourmia melon virus
	Pelargonium zonate spot virus

Table 4.4. Satellites and Viroids.

Type	*Genome*	*Subgroup*	*Type species*
Satellites	ssRNA	Subgroup 2A	Tobacco necrosis
	Satellite RNA	Subgroup 2B Type	satellite
		mRNA Satellites	
		Subgroup 3C Type	
		linear RNA	
		Satellites Subgroup	
		4D Type	
		circular RNA	
		Satellites	
Viroids			Potato spindle tuber viroid

Recently, however, there has come a move to introduce higher-level taxa such as the "supergroup" or family into plant virus classification schemes, in line with animal and human virus classification. There has also been an increasing use of genome sequence data for establishing relationships between viruses: some of these analyses have been especially illuminating in that they have revealed distant relationships between several plant virus groups and some animal virus families.

Two notable examples are the relationships between picornaviruses, comoviruses and potyviruses; and between caulimoviruses, retroviruses and hepadnaviruses.

Such sequence comparisons have largely made use of strongly-conserved predicted protein sequences that are related to replication; the inference of the analyses is that the replicases of individual viruses within these separate groups of viruses have diverged evolutionarily from a common source.

Similar sequence comparisons have been used to estimate evolutionary distances between, and make taxonomic proposals for, many different organisms, ranging from bacteria to plants to hominoids. A common approach in these studies has been the construction of phylogenetic trees for the organisms in question: these have been used to clarify lines of evolutionary descent of the organisms, and illustrate relationships with taxonomic significance. It is interesting that very few such analyses have been performed for viruses in general, and plant viruses in particular - probably both because of a general reluctance to speculate upon virus evolutionary processes, and because of a lack of sequence data. The analyses that have been done have also concentrated mainly on distance data - that is, pairwise similarity or difference estimates - for use in phenetic comparisons, rather than using the cladistic techniques which have become popular in animal and plant systematics recently.

Data Sets Used for Relationship Dendrograms and Phylogenetic Analyses

Use of serological data

Some of the first quantitative estimates of "difference" between plant viruses were based on serological data. Serology has long been used for grouping plant viruses. However, the concept of quantitating serological differences is fairly recent. In 1970 van Regenmortel and von Wechmar introduced the concept of the serological differentiation index or SDI: this is the serological cross-reactivity between two viruses expressed as the number of twofold dilution steps separating homologous and heterologous titers.

Formerly such investigations were performed using precipitin techniques; recently, however, ELISA techniques have been used for SDI determinations: Jaegle and van Regenmortel demonstrated the general use of the technique, and Dekker et al. (1988) and Pinner et al. (1992) put it to use in determining relationships among maize streak virus (MSV) and related geminiviruses of cereals and grasses.

SDI data have been used for taxonomic purposes, in that "clusters" of related viruses can be differentiated from less-related species. It has also been used to construct relationship dendrograms: Koenig (1976) presented a comprehensive analysis of the serology of 13 tymoviruses, which included a novel "loop structure".

In fact, one can take a table of reciprocal SDI values for a group of viruses and analyse it directly for construction of a dendrogram. The SDI data is conceptually similar to the pairwise DNA hybridisation data that has often been used for construction of phylogenetic trees); thus, one may also root the dendrogram to make it into a tree, and speculate on the phylogeny of virus coat proteins from serological data.

Coat protein composition

Some virologists have shown that the relatedness of tobamovirus coat proteins as assessed from amino acid sequences agrees closely with that assessed from their amino acid compositions. These latter data to construct relationship dendrograms for many tobamoviruses: these dendrograms may also be regarded as phylogenetic trees, as they are rooted, and reflect a line of evolutionary descent. These may productively be used for re-classification of certain of the viruses, and as confirmation of the classification of several others: for example, closely-clustered viruses could be regarded as strains, and distinct clusters as groups of strains of different viruses.

That the approach may be comfortably used to define a taxonomic group - or genus - is shown by Gibbs (1980), where a tree that includes the furovirus beet necrotic yellow vein virus (BNYVV) clearly separates this from definitive tobamoviruses.

Fauquet et al. (1985) have also claimed that the amino acid compositions of most virus coat proteins fell into groups closely corresponding to accepted plant virus taxonomic groups: thus determination of the composition of a virus coat protein may well be a most useful exercise from a taxonomic point of view, as such data appears to define the virus taxonomic group. However, it is still an analytical undertaking of some magnitude, and difficult for most laboratories.

Limited RNAse digestion and oligonucleotide mapping

For many years researchers have been mapping and typing ssRNA viruses of mammals (and of plants) by limited RNAse digestion of genomic RNA and oligonucleotide mapping by electrophoresis and chromatography. Such analyses are obviously possible for plant viruses, and could be as informative. They are only really usable at the level of "species" relationships, as distantly-related RNAs tend to have very different oligonucleotide maps.

Restriction fragment and map data

It is a relatively simple task - and certainly easier than exhaustive protein purification protocols - to purify double-stranded DNAs from plants, and to analyse them by restriction endonuclease digestion, electrophoresis, and hybridisation with labelled probes. The use of restriction endonuclease cleavage patterns and derived restriction maps of DNA sequences for phylogenetic analysis - particularly of mitochondrial DNAs - is well established; however, the methods have had little application in plant virology since most plant viruses have ssRNA genomes.

One group of viruses that have been studied by restriction mapping and restriction fragment pattern differences are the geminiviruses of maize and cereals.

It is possible to directly convert a series of restriction fragment patterns of different viral DNAs on an electropherogram into a digital data matrix and analyse this by cladistic techniques; it is also possible to construct difference tables and analyse them phenetically. Detailed restriction maps of related virus DNAs may be transformed into distance matrices or also treated as digital information for cladistic analysis.

With recent developments in cDNA synthesis technology, it is also possible to routinely map RNA viruses by restriction enzyme cleavage: for instance, DNA fragments obtained from human rhinoviruses by polymerase chain reaction (PCR) amplification of cDNA have been used for typing of the virus isolates by restriction endonuclease cleavage patterns; PCR has also been used for amplification and typing of viroids from cDNA. Restriction map data falls down, however, at demonstrating relationships beyond the level of about 30% sequence difference. Thus map comparisons may only be useful for sub-group level comparisons.

Nucleic acid and protein sequences

Recent developments in nucleic acid technology make it almost easier to obtain long stretches of RNA or DNA sequence, than to sequence or to determine the composition of the protein coded for by the sequence; however, the latter has also become easier (Shukla and Ward, 1989). Thus, comparisons of coat protein gene sequences - or of the predicted amino acid sequences coded for - are probably going to be far more widely used for comparison purposes than are serological data or amino acid composition data. The usefulness of coat protein sequence differences for phylogenetic analysis of the tobamoviruses has already been mentioned; it is worth noting that Gibbs (1976) has used such an analysis to estimate a divergence time of between 200 and 600 Myr for the definitive tobamoviruses. Both sorts of analysis - directly from sequence, and from pairwise sequence difference comparisons - lead to diagrams which can be used to group viruses.

Application of Phylogenetic Analyses to Plant Virus Classification

A convenient rule of thumb with the simpler viruses appears to be that coat protein gene sequences vary the most, followed by certain of the non-structural proteins such as movement proteins, followed by proteins associated with the replication machinery. This is nowhere more clear than among the plant virus groups linked by possession of a tripartite ssRNA genome and isometric or bacilliform particles: the bromo-, cucumo-, ilar- and alfalfa mosaic virus groups, which have been proposed as members of a plant virus family to be called the Bromoviridae.

These examples could be taken as leading inexorably to a proposal for the establishment of plant virus taxa right up to the "super-familial" level. However, plant virus genome relationships are not as simple as are the phylogenetic relationships of some of their constituent parts and any taxon higher than the familial level would be very hard to justify. For example, the Bromoviridae all have a similar genomic structure and organisation, and it could be confidently asserted that all components of all of them have a common source. Tobamoviruses also all have a similar genetic organisation. Their putative replicases are related to those of the Bromoviridae; however, there is no demonstrable relationship between any other genes of the two groups of viruses, and the genetic organisation is markedly different. Moreover, both sets of viruses

Table 4.5. Classification of plant virus based on phylogenetic analysis.

Organism (by category)	*Strain*	*Common Name*	*Genome Size*	*Rationale or Significance*
Potyvirus	MDMV WSMV	Maize dwarf mosaic Wheat streak mosaic	10.4 kb	Largest single taxonomic group of plant viruses, transmission by aphids (nonpersistentmanner), whitefly or fungi, some seedborne, include leguminous, solanaceous and monocot hosts, global distribution.
Tritimovirus	PVY BCMV SMV	Potato Y Bean common mosaic Soybean mosaic		
Closteroviruses /Criniviruses	CTV ToCV	Citrus tristeza Tomato chlorosis	17-20 kb/15kb	Viruses of woody and herbaceous hosts, large complex genomes of ssRNA, semi-persistent,
	PMWV	Pineapple mealy bug		Associated with hemipteran vectors (aphid, mealy bug and whitefly).
Begomoviruses	SLCV BGMV AbMV	Diverse strains and species: Bipartite New World or Old World monopartite	5.2 kb	Only circular ssDNA plant viruses, global, greater than expected variability, apparent rapid evolution through recombination and reassortment, poor association between
Curtoviruses	PHV clades BCTV and	viruses of cultivated and wild species	2.8 kb	viruses of cultivated and weed hosts, several hemipteran vectors

Continue...

	related viruses			
Tospoviruses	TSWV	Tomato spotted wilt	~17 kb	Large, complex, ssRNA virus replicates in insect vector. Medical significance (Bunyaviridae).
Tombusvirus	TBSV	Tomato bushy stunt	4.8 kb	Highly stable, ssRNA viruses transmitted in persistent manner by fungal zoospores
	CNV	Cucumber necrosis		
Luteovirus/ Polerovirus	BYDV	Barley yellow dwarf	5.5 kb	ssRNA viruses transmitted in persistent, circulative manner by aphids, worldwide distribution, infect monocots and dicots.
	BWYV	Beet western yellows		
Cucumovirus	CMV stains	Cucumber mosaic virus	8.0 kb	One of the largest groups of plant viruses, have broad host ranges, but poorly understood biodiversity. Global distribution.
Rhabdovirus	SYVV	Sonchus yellow vein virus	13.0 kb	Large, complex, membrane bound RNA viruses, replicates in insect, medical significance through relationship to animal viruses.

share sequence similarity with the animal alphaviruses, in the family Togaviridae, which have even less similar genetic organisation.

Much the same can be said for the picorna-, como- and potyviruses. Some virologists have proposed that simple viruses have evolved by building up modules into genomes, and that different "supergroups" can be distinguished by their core modules, which consist of genes related to replication. Thus there is a picornavirus-like supergroup of viruses with similar polymerse/ protease core modules and an alphavirus supergroup of viruses with similar "nucleotide-binding" protein/replicase cores.

With plant viruses, the term specificity (or host-specificity) has a very narrow meaning, since no plant virus as such exists. Instead, plant viruses can be grouped in a number of 'varieties'. The tobacco mosaic-virus (TMV), for example, multiplies within Nicotiana-species, several other solanaceous plants, and a few species of other plant families. The name of a virus is usually derived from the name of its main host plant. Although with viruses, the term species may not quite correspond to the way it is defined in biological systematics, it is perfectly reasonable and common to use it for viruses, too, since all viruses and viroids contain an original genome with a species-specific information. Its continuity over generations is guaranteed by replication in the host cells. The genetic information of viruses is either encoded by single-stranded RNA (most plant viruses), double-stranded RNA (wound tumor viruses), single-stranded DNA (gemini-viruses) or double-stranded DNA (cauliflower mosaic-virus: CaMV). Based on the shape of the virus particle, it is distinguished between rod-shaped and icosaedrical viruses with a capsid that seems almost spherical.

Plant viruses have no specific mechanism for entering the host cell. Cell wall and cuticle are difficult obstacles for them. Plant viruses depend therefore on injuries or on transmission via invertebrates (insects, nematodes, etc.). The animal transmitter does in some cases also act as an intermediate host. This means that some plant viruses are able to multiply within animal tissue.

Virus diseases of plants are relatively rare. Infection is scarcely strong enough to kill the plant. Monoculture favours spreading, and agricultural losses of profit caused by the potato-X, or the potato-Y viruses, for example, can be significant.

Of many viruses, numerous, considerably differing strains (wildtypes) have been isolated. Among the differences are host range and the degree of virulence. Moreover, viroids were detected. Viroids are small, circular RNA molecules that do not encode proteins themselves. Instead, virions interfere with the transcription of cells due to their similarity with certain areas of recognition of primary transcription products. It seems that viroids prevent the correct cutting out of the introns. They are presumably multiplied with the aid of the cellular DNA-dependent RNA-polymerase II. Viroids occur mainly in warm climates and cause significant loss of profit as the causative agents of the potato disease or the Cadang-Cadang disease of palms.

The virus concentration within plant cells is high, although a virus like the TMV does not harm the host seriously. Infected cells contain often voluminous virus crystals. TMV was thus the first virus to be isolated in a pure state and in large amounts.

Plants are far from being defenceless against viruses. Only a few virus species are able to penetrate meristematic tissues or to infect a number of successive plant generations (vertical transmission). Hypersensitivity is an effective protection. It is based on the dying of cells in the immediate surrounding of the primary site of infection, thus stopping the spreading of the virus. In the case of Nicotiana tabacum, the genetic cause of hypersensitivity has been analyzed. One Nicotiana gene product guarantees hypersensitivity against all TMV strains, while another affects only some TMV strains. The symptom a virus causes at the primary site of infection is called primary symptoms. Symptoms caused by its spreading throughout the rest of the plant are called secondary symptoms.

Virus infections can usually be recognized by mosaic-like leaf patterns of light and dark green. The infection spreads often over the whole leaf beginning at the leaf veins. Leaves that had been infected during their development are usually deformed or involute.

Frequently, lightened leaf areas, called chloroses, develop around the primary site of infection. Withered areas are called necroses. Chloroses are caused by a breakdown of the chlorophyll resulting in a decreased rate of photosynthesis. Heavy infections are characterized by a complete local loss of chlorophyll. Affected areas have a yellowish look as only the carotenoids remain. Some TMV strains, for example, can be recognized by yellow leaf areas ("yellow strains"). In nature, such strains are pretty rare since such a strong damaging of the host does also impair the virus' chances of replication and spreading.

Some viruses multiply within the plant without causing symptoms. This phenomenon is called leten infection. In contrast, wound tumor virus causes the development of tumors. The symptoms of most viruses are dependent on both virus and host, and do thus present an important diagnostic feature.

G. Samuel proved in 1934 experimentally that the TMV virus uses the plant's vascular system for spreading, actually both the plant's transport to and way. As a result, fully differentiated, old leaves and roots, and young leaves are equally infected.

Since decades now, plant viruses (and especially the TMV) are favourite test objects of pure research due to the large amounts of material extractable from infected plants. The TMV was, too, the first biological electron-microscopic object to be photographed.

Isolated TMV-RNA itself is infectious. It can thus be a carrier of genetic information. Chemical changes of the TMV-RNA (deamination of adenine or cytosine due to treatment with nitrite) showed in addition and for the first time that changes of the RNA are mutagenic, too. The changes of these first experiments caused mutations of a coat protein.

The analysis of a large number of chemically induced mutants (performed by the working groups above yielded the first results important for the encoding of the genetic code. Work on the TMV made plain, too, that the gene-carrying RNA does not simultaneously function as mRNA. mRNA is synthesized within the cell as a separate fraction.

Table 4.6. Plant viruses Carlavirus.

Name	*Accession Number*
Carnation latent virus	DSMZ PV-0365
Carnation latent virus	DSMZ PV-0366
Chrysanthemum virus B	DSMZ PV-0458
Cowpea mild mottle virus	DSMZ PV-0090
Garlic common latent virus	DSMZ PV-0449
Helenium virus S	DSMZ PV-0336
Kalanchoe latent virus	DSMZ PV-0290
Passiflora latent virus	DSMZ PV-0222
Poplar mosaic virus	DSMZ PV-0341
Potato virus M	DSMZ PV-0273
Shallot latent virus	DSMZ PV-0425
Shallot latent virus	DSMZ PV-0426

Table 4.7. Plant viruses Tymovirus.

Name	*Accession Number*
Belladonna mottle virus	DSMZ PV-0042
Eggplant mosaic virus	DSMZ PV-0060
Eggplant mosaic virus	DSMZ PV-0061
Eggplant mosaic virus	DSMZ PV-0062
Okra mosaic virus	DSMZ PV-0264
Poinsettia mosaic virus	DSMZ PV-0340
Turnip yellow mosaic virus	DSMZ PV-0299

Table 4.8. Plant viruses Caulimovirus.

Name	*Accession Number*
Cauliflower mosaic virus	DSMZ PV-0226
Cauliflower mosaic virus	DSMZ PV-0227
Cauliflower mosaic virus	DSMZ PV-0228
Cauliflower mosaic virus	DSMZ PV-0229
Cauliflower mosaic virus	DSMZ PV-0362

Table 4.9. Plant viruses Potexvirus.

Name	*Accession Number*
"Tamus red mosaic virus"	DSMZ PV-0397
Cactus virus X	DSMZ PV-0083
Cymbidium mosaic virus	DSMZ PV-0334
Hydrangea ringspot virus	DSMZ PV-0223
Hydrangea ringspot virus	DSMZ PV-0246
Hydrangea ringspot virus	DSMZ PV-0372
Potato aucuba mosaic virus	DSMZ PV-0007
Potato virus X	DSMZ PV-0014
Potato virus X	DSMZ PV-0015
Potato virus X	DSMZ PV-0017
Potato virus X	DSMZ PV-0018
Potato virus X	DSMZ PV-0019
Potato virus X	DSMZ PV-0020
Viola mottle virus	DSMZ PV-0224
White clover mosaic virus	DSMZ PV-0310

Table 4.10. Plant viruses Cucumovirus.

Name	*Accession Number*
Cucumber mosaic virus	DSMZ PV-0035
Cucumber mosaic virus	DSMZ PV-0036
Cucumber mosaic virus	DSMZ PV-0037
Cucumber mosaic virus	DSMZ PV-0038
Cucumber mosaic virus	DSMZ PV-0183
Cucumber mosaic virus	DSMZ PV-0184
Cucumber mosaic virus	DSMZ PV-0185
Cucumber mosaic virus	DSMZ PV-0187
Cucumber mosaic virus	DSMZ PV-0188
Cucumber mosaic virus	DSMZ PV-0189
Cucumber mosaic virus	DSMZ PV-0314
Cucumber mosaic virus	DSMZ PV-0357
Cucumber mosaic virus	DSMZ PV-0358
Cucumber mosaic virus	DSMZ PV-0359
Cucumber mosaic virus	DSMZ PV-0418
Cucumber mosaic virus	DSMZ PV-0419
Cucumber mosaic virus	DSMZ PV-0434
Cucumber mosaic virus	DSMZ PV-0445
Cucumber mosaic virus	DSMZ PV-0453
Peanut stunt virus	DSMZ PV-0190
Peanut stunt virus	DSMZ PV-0203
Peanut stunt virus	DSMZ PV-0384
Tomato aspermy virus	DSMZ PV-0068
Tomato aspermy virus	DSMZ PV-0312

Table 4.11. Plant viruses Comovirus

Name	*Accession Number*
Andean potato mottle virus	DSMZ PV-0057
Andean potato mottle virus	DSMZ PV-0058
Andean potato mottle virus	DSMZ PV-0059
Broad bean true mosaic virus	DSMZ PV-0098
Cowpea mosaic virus	DSMZ PV-0074
Cowpea severe mosaic virus	DSMZ PV-0050
Radish mosaic virus	DSMZ PV-0306
Radish mosaic virus	DSMZ PV-0355

Table 4.12. Plant viruses Tombusvirus

Name	*Accession Number*
"Pelargonium virus, unnamed"	DSMZ PV-0371
Carnation Italian ringspot virus	DSMZ PV-0069
Cucumber necrosis virus	DSMZ PV-0313
Cymbidium ringspot virus	DSMZ PV-0272
Neckar river virus	DSMZ PV-0270
Pelargonium leaf curl virus	DSMZ PV-0265
Pelargonium leaf curl virus	DSMZ PV-0370
Petunia asteroid mosaic virus	DSMZ PV-0339
Tobacco necrosis virus + satellite	DSMZ PV-0198
Tobacco necrosis virus + satellite	DSMZ PV-0218
Tomato bushy stunt virus	DSMZ PV-0268
Tomato bushy stunt virus	DSMZ PV-0269
Tomato bushy stunt virus	DSMZ PV-0284
Tomato bushy stunt virus	DSMZ PV-0285

Double-Stranded RNA-Viruses; Wound Tumor Viruses:

The genome of wound tumor viruses (WTV) consists of 12 double-stranded RNA segments. Neither of them is infectious. The wound tumor viruses of plants and animals (reoviruses) are related and are characterized by a number of common activities. They contain, for example, the enzyme transcriptase that transcribes single-stranded RNA or, in other words, produces an mRNA complementary to the transcribed strand. Each of the 12 segments can be transcribed and it is assumed that each of them encodes one protein. Replication occurs within the cytoplasm. Infection takes place via insects (e.g. aphids) that function both as a vector and an intermediate host, i.e. the virus multiplies in their tissue, too.

More than 50 plant species are known to be susceptible for wound tumor viruses. Among the symptoms are small tumors at the stem and larger and more numerous ones at the roots. WTV-induced wound tumors of leguminosae can be clearly distinguished from the nodules caused by nodule bacteria. In some plant species, like the lobelia, the infection induces the development of organs from otherwise normal organs, e.g. a leaf can develop at the lower leaf surface of another leaf.

Plant Viruses with Circular, Single-Stranded DNA: Gemini Viruses:

The particles of gemini viruses are quasi-isometric. They are called gemini (twin) viruses, because they are usually found in pairs. Each particle has a diameter of just 15 - 20 nm. Gemini viruses belong to the smallest virus particles able to multiply without a helper virus. They have a circular DNA with a molecular weight of 0.7 - 0.8 x 10^6 (about 2,500 base pairs). In the case of some gemini viruses, it has been proven, in the case of others, it is assumed, that the genome consists of two molecules of DNA of almost equal size, but different sequence. The nucleotide sequences of some species are known.

Isolated circular DNA alone is not infectious. In infected host cells, the nucleus holds the chief amount of viral DNA. It is therefore assumed that the nucleus is also the place of its replication. It looks as if - analogous to the TMV - double-stranded intermediates would exist (= replication state: RF), since double-stranded DNA has been isolated from infected cells. Gemini viruses are of interest to genetic engineers, since they can infect monocots where their DNA enters the nucleus, too, and since the interest in monocot vectors is still large. The bean golden mosaic-virus (BGMV), the cassava latend-virus (CLV), the tomato golden-mosaic-virus (TGMV), the maize-streak virus (MSV), and the abutilon mosaic-virus belong all to the Gemini-virus family.

Fig. 4.6. Viral infection in leaf.

Insects (greenhouse whitefly, grasshoppers, and others) help usually in spreading Gemini-viruses in nature. These viruses can cause considerable damage to agriculture.

Double-Stranded DNA Viruses

The prototype of a plant virus with double-stranded DNA is the cauliflower mosaic-virus (CaMV). Double-stranded plant viruses, too, have been considered in genetic engineering. It is tried to convert them into vectors in order to put foreign DNA into plant cells similar to the way the plasmid of Agrobacterium tumefaciens is used.

CaMV-DNA is normally not incorporated constantly into the host cell genome (R. Sphepherd and R.J. Wakeman, 1971). CaMV is the collective name for a group of tightly related viral species usually transmitted by aphids. Each species has a narrow host specificity, overlapping with that of other species is rare. The virion is spherical with a diameter of about 50 nm. Most likely, the capsid consists of 420 identical subunits with molecular weights of 42,000. The DNA contains about 8,000 base pairs. In the case of one species, it has been sequenced and 8,024 base pairs were found. Neutron diffraction experiments showed that the DNA is sandwiched between the protein subunits. The core of the viral particle contains no proteins or nucleic acids. One of the three ring-shaped DNA molecules is broken up in three different places. Treatment of the viral DNA with SI-endonuclease, a DNAse breaking down single-stranded DNA, renders therefore three fragments of defined length. Further analyses showed that the ring is no closed, uniform structure, but that it results from the combining of three molecules kept together at their ends by complementary sequences.

Symptoms occur two to three weeks after infection and can be recognized by the mosaic-like lesions of infected leaves. The virus spreads systemically, its secondary symptoms are similar to those of the primary infection. Leaves that were infected during their development display deformed leaf blades.

Viroids[*], the Smallest Infectious Units

Viroids are infectious units that cause a number of plant diseases. They are circular molecules of RNA with molecular weights between 107,000 and 127,000. H. J. Gross et al. sequenced the nucleotide sequence of the potato spindle tuber virus (PSTV) in 1978. It consists of 359 ribonucleotides and is characterized by numerous intramolecular base-pairings that lend stability to the structure. They are organized in a sequence of helices separated from each other by loops. The resulting structure resembles a dumb-bell with an axis ratio of 1:20. Several more viruses have been sequenced in the meantime. All of them have structures similar to that of the PSTV. They are ~240 - 380 nucleotides long and all of them have dumb-bell structures. The fact that a central portion of the molecule that is responsible for the pathogenicity of the viroids, is structurally conserved is especially interesting. PSTV-mutants with changes in this section are less pathogen than the wild type (M. Schnolzer et al., 1985).

Viroids multiply even at relatively high temperature (about 35°C). Most likely, they have adapted to their host plants that have so-far strictly been found to inhabit tropical, subtropical, and continental climates. The viroids are localized within the chromatin fraction of the nucleus. The DNA-dependent RNA-polymerase II and I use the viroids as templates and produce strands that again serve as templates for the synthesis of the +strand (E. Spiesmacher et al., 1985).

5

Plant Viral Diseases

People have been combating weeds, insects, and plant diseases throughout history. Pests of the plants man use for food are abundant. Pests (insects and weeds), and diseases, developed "hand in hand" with the plants they attacked. Pests are plants, animals or viruses that are detrimental to humans and crops. Cultivation of one or a few crops species on the same land for several years or decades shifts the population balance of soil microorganisms, plant pathogens, insect pests, and weeds to those strains favored by the crops grown. Most crops are subject to attack from a large number of pests (insects, weeds) and diseases. These are considered to be the major limiting factor in crop production. The susceptibility of cultivars to attack from specific pests or diseases varies greatly. Development of resistant lines would seem to offer the best prospects of crop protection, particularly if this is combined with the use of healthy, clean seed, efficient weed control and crop rotation. It is very difficult to make reliable estimates of losses due to pests. There is no question, however, that these losses are of substantial and often staggering proportions. A conservative estimate suggested that plant pests destroyed one third of mankind's supply of food and fiber every year. Losses of much greater magnitude occur as a matter of course in many less developed countries. It is only in recent times that man's response to pest control has been based on any appreciable understanding of the nature and causes of pest problems. Methods of pest control change and become increasingly effective as we gain greater understanding of pests and their habits. Although there remains much that we do not know, we can formulate pest control programs on a rational basis. A control program should be based on an understanding of the biology and habits of the pest, a consideration of all effective methods of control, and recognition of the level of control that is both desirable and possible. Some of the preventive measures and control methods for pests and diseases are discussed in this lesson.

Plants and plant products are tremendously important for human survival since they provide food, clothing, furniture, a stable environment, and often housing. Plants, whether cultivated or wild, generally grow well when the soil provides them with sufficient nutrients and moisture, sufficient light reaches their leaves, and the temperature stays within a "normal" range. However, like people, plants can get sick. Agents similar to those that cause disease in people also can cause diseases in plants. The broadest definition of plant disease includes anything that damages plant health. This definition can include such diverse factors as pathogens, insufficient nitrogen, air pollution, lawnmower damage, and deer damage. A stricter definition usually includes any persistent irritation resulting in plant damage and characteristic symptoms. This definition includes such factors as pathogens, insufficient nitrogen, and air pollution.. However, it excludes factors such as lawnmower injury to trees and lightning injury since this damage presumably is a one-time occurrence. A very strict definition includes only infectious organisms (pathogens) that multiply and spread to other nearby plants. Most pathogens are microscopic and include bacteria, fungi, nematodes, viruses, mollicutes, protozoa, and parasitic plants. These are called biotic diseases. Plant pathologists usually do not consider organisms such as insects, deer, rodents, and birds to be pathogens.

As a science, plant pathology strives to increase our knowledge about plant diseases. To do this, plant pathologists study such factors as the biology of the pathogenic organisms, the mechanisms that pathogens use to cause disease and interactions between pathogens and host plants. These studies help plant pathologists to devise better methods of preventing or controlling diseases and alleviating the damage that they cause.

In order for a biotic plant disease to occur, three conditions must be met:

1. The host plant must be susceptible.
2. A pathogen must be present.

3. The environment must be favorable for infection by the pathogen.

All three of these factors must occur simultaneously. If one or more is absent, the disease does not occur. The genetics of the plant determine its susceptibility or resistance to a particular disease. Specific pathogens vary in their ability to infect different plant species. Various physical and biochemical factors of a given plant species influence susceptibility. These factors include plant defense mechanisms, carbohydrates and protein types, cuticle thickness, and stomatal shape. The developmental stage of the plant also can influence disease development. Pathogens differ in their ability to survive, spread, and reproduce. For instance, a certain viral disease may occur only when a specific insect vector transports virus particles to the susceptible plant. Certain temperature, light, or moisture conditions generally are necessary for a disease to occur.

Since the first time modern plant breeding began, incorporation of genes conferring pathogen resistance into commercial varieties has been a goal of importance. To get satisfactory result requires that in the plant genomes, resistance genes exist. It also requires that the resistance genes be compatible with the crop variety being developed. Once introduced by a series of crosses the resistance genes must be effective and have good resistance to the target problem. For example, the N gene, which was resistant to tobacco, fifty years ago from a tobacco relative, Nicotiana glutinosa and this virus-resistance is still widely used effectively. Most of the time however, resistance doesn't last. The population of the viral stains, where there is a race between plant breeders using new resistance genes and the pathogens that are targeted, overcomes resistance. Many crop plants are affected by viral disease, for which there is no source of resistance in compatible plant genotypes. A lot of people are unaware of the impact of plant viruses. All over the world serious problems from plant viral diseases come up. That's why, when the first virus-resistant transgenic plants were described, the workers in the field were definitely excited. A major ecological risk associated with the widespread cultivation of transgenic crops is genetic flow of the transgene to non-transgenic populations of the same, or other, species. If genetic flow to viral population could occur in nature, and fit recombinant viruses could spread in the field. Weather this possibility is more probable than or as probable as that of similar recombinant viruses originating from mixed infections in non-transgenic plants is a point that needs to be evaluated. Recombinant viruses can sometimes be hard to fight. Geminiviruses transmitted by the white fly (begomoviruses) is a good example of emerging viruses in the plant kingdom. Virus groups have analyzed the emergence and in some instances recombination between the pre-existing white flies and new ones seem to be at the root of the appearance of new viruses and stains that cause devastating epidemics. Infecting viruses are a big concern since they could lead to recombinant viruses with modified biological properties. RNA recombination occurs naturally in viral genomes, which there is clear evidence to prove it. It is generally considered that, mutation, assortment, and recombination are three main molecular forces that increase variability of the revolution of RNA viruses. Plant viral disease dangers can be devastating to farmers, ranchers, and the surrounding communities.

Viruses are so small that they cannot be seen with an ordinary microscope. Their effects on plants generally recognize them. Often it is difficult to distinguish between diseases caused by viruses or mycoplasmas and those caused by other plant disease agents such as fungi and bacteria.

Usually, the best way to identify a virus is to compare the symptoms with pictures and descriptions of diseased plants for which a positive identification has been made. Other methods require more sophisticated testing, such as inoculating indicator plants and observing the results or using specifically identified antibodies to test for the presence of the organism.

Viruses depend on other living organisms for food and to reproduce. They cannot exist separately from the host for very long. Viruses are commonly spread from plant to plant by mites and by aphids, leafhoppers, whiteflies, and other plant- feeding insects. They may be carried along with nematodes, fungus spores, and pollen, and may be spread by people through cultivation practices, such as pruning and grafting. A few are spread in the seeds of the infected plant.

Viruses can induce a wide variety of responses in host plants. Most often, they stunt plant growth and/or alter the plant's natural colour. Viruses can cause abnormal formation of many parts of an infected plant, including the roots, stems, leaves, and fruit.

Viruses usually cause Mosaic diseases, with their characteristic light and dark blotchy patterning. Viroids are similar to viruses in many ways, but they are even smaller and lack the outer layer of protein that viruses have. Only a few plant diseases are known to be viroid-caused, but viroids are the suspected cause of many other plant and animal disorders. Viroids are spread mostly through infected plant stock. People can spread infected plant sap during plant propagation and other cultural practices. A few viroids are known to be transmitted with pollen and seeds. Mycoplasmas are the smallest known independently living organisms. They can reproduce and exist apart from other living organisms. They obtain their food from plants. Yellows diseases and some stunts are caused by mycoplasmas. Insects spread most mycoplasmas, most commonly by leafhoppers. Mites may also spread them. Mycoplasmas are also readily spread among woody plants by grafting.

As discussed in previous chapters, virus is an extremely small organism composed of genetic material (either RNA or DNA) surrounded by a protein coat. Viruses replicate only inside living cells. Viruses are able to take command of plant cell metabolism and direct the cell to manufacture more virus particles. An average virus is 100,000 to 1,000,000 times smaller than a plant cell. Due to their size and biology, plant viruses traditionally have been difficult to identify. Symptoms give important clues, but yield no real proof of the exact pathogen involved. In recent years, many advances in molecular biology techniques have allowed plant pathologists to "label" viruses in order to identify them. Virologists have developed many different labeling techniques for virus identification. There is some example of one such technique. For detecting tobacco mosaic virus (TMV) in tomato, virologists have developed a substance that turns blue only when a solution of crushed plant material containing TMV is present. If the tomato spotted wilt virus (TSWV) is present instead, the solution remains clear. To confirm the presence of TSWV, one must conduct a second test containing a substance that turns color in the presence of TSWV. It would be too time consuming to run tests for the 30 possible tomato viral diseases. This is where the art of plant disease diagnosis is important. A skilled diagnostician can look at the plant symptoms, make guesses as to the viruses present, and then run tests specific for these viruses to scientifically identify the disease. Through previous chapters, it is cleared that viroid is a piece of genetic material that, unlike a virus, has no protein coat. Viroids divert plant metabolism to produce more viroids. They spread by vegetative propagation.

Current Strategies Used to Control Plant Viruses

1. The principle method is to identify resistance genes in a species and incorporate these genes into cultivars and varieties that have other good agronomic or horticultural properties. In some cases there are single genes that provide resistance to a particular virus. In others, there may be a small number of genes that in combination provide resistance. These are known as host resistance genes. In the last two years, the first of these genes for resistance to a virus, in this case against tobacco mosaic virus, has been cloned.
2. A second method to control viruses is by controlling how they are spread. For viruses that are spread mechanically, that means minimizing contact between healthy and infected plants, and anything that is shared between them, such as greenhouse pruning tools. Eliminating smoking in greenhouses is apparently important to controlling the spread of TMV, which is present in cigarette tobacco.
3. For viruses transmitted by vectors, controlling the population of the vector is another method of control. This can include cultural practices. For example, it is recommended that growers delay planting winter wheat until after the "fly free day". Not only does this reduce the damage that can be caused by the Hessian fly, it also prevents infection of the wheat seedlings with Barley Yellow Dwarf Virus (BYDV), which is transmitted by aphids. Aphids are no longer active after the "fly free day" and so cannot infect the crop in the fall. The wheat can still be infected by BYDV in the spring, but a later infection is less likely to cause serious losses.
4. Another method to control virus infections in some crops is called cross-protection. This is used in tomatoes and citrus crops. Plants are deliberately infected with a mild strain of a virus, which produces only mild symptoms. The plants that have been infected with the mild strain of the virus are in some way protected against a subsequent infection by a more severe strain of the virus. Plants do not have an immune system similar to that of mammals and this is not the same as a vaccination that animals might have received but the outcome

is similar. The mild strain of virus protects the plant from a subsequent infection. This formed the basis for experiments to examine this phenomenon and led to the development of a general method to make plants resistant to viruses. The hypothesis was put forward that some component of the virus was responsible for triggering this resistance phenomenon.

The Life Cycle of a Typical Plant Virus

1. When a virus enters a cell, the RNA is released from the protein coat.
2. This RNA is then used as template to make RNA and proteins encoded by the virus. The virus uses the machinery of the cell to express the genetic information carried by the viral genome.
3. The genome of the virus is replicated and new coat proteins are synthesized. This leads to the assembly of new virus particles in the infected cells.
4. Finally the virus is transmitted to another plant, and this can be done in a number of ways:
   (a) Mechanically by physical touching of infected and healthy plants; for example, in greenhouses some viruses are spread by using the same tools, such as pruners, which carry virus from infected plants to those that are healthy. Good sanitary practices can reduce the spread of viruses by pruning tools.
   (b) By a biological vector, such as aphids or other insects that feed on an infected plant, ingest some virus particles, then move to a healthy plant and transmit the virus in the saliva when feeding.
   (c) Some viruses are carried in the seed and so are passed from one generation to another

Symptoms and Viral Pathogenesis

Type of Viral Symptoms

Symptoms of viral diseases were well described even before viruses were well characterized. In early days when techniques for virus purification and visualization were not available, virus detection and identification were entirely based on symptoms. For this historic reason, most virus names were derived from the symptoms they cause on host plants. In most cases, you can tell typical symptoms of a virus from its name. For example, tobacco mosaic virus (TMV) obviously cause mosaic symptom on tobacco plants.

Virus infection often alters the host metabolism and interrupts important cellular functions. The effect of these alterations is expressed macroscopically as deviations from normal morphology. When a plant does not support virus replication or translocation, no viral symptoms are induced. These plants are called resistant plants. Susceptible plants are the plants that allow infection and movement of viruses.

A combination of several symptom types usually shows up in a plant infected by a virus. For example, wheat or barley plants infected by barley yellow dwarf virus are dwarfed and all the leaves turn yellow in color. For the following description, we will consider each symptom type individually.

No obvious symptoms

Not all virus infections would necessarily cause visible symptoms on infected plants. A virus can multiply to a very high concentration and yet induce no obvious symptoms in an infected plant. Infection with a mild strain of the virus in a tolerant host will result in a symptomless infection. A classical example is TMV on Ambalema tobacco. TMV infects and replicate throughout tobacco plants, but infected plants show no signs of disease. A more practical example is citrus tristeza virus (CTV). Citrus tristeza virus causes quick death of citrus on sour orange rootstock. In Meyer lemon tree, CTV replicates to a high titer and the tree remain normal in appearance. Plants that are infected with viruses but do not display symptoms are called symptomless carriers. Symptomless carriers play a significant role in a virus epidemic. Two genera from the family of Partiviridae: alphacryptovirus and betacryptovirus infect plants but do not induce obvious symptoms on infected plants. This group of viruses have double-stranded RNA genome. They are transmitted efficiently though pollen or seeds but are not transmissible mechanically or by vectors. These viruses may have co-evolved with their hosts and become almost a part of the host genetic materials.

Localized symptoms (local lesions)

Local lesions usually develop near the site of entry on the leaves or fruit. Most types of symptoms see are chlorotic or necrotic spots, ringspots, ranging from single rings from concentric rings of rings of chlorotic or necrotic nature. Local symptoms are usually a result of interaction between host defense responses and virus infections. These defense responses include rapid death of cells (hypersensitivity) and other physiological changes, which limit virus spread further. Viruses are limited to local lesions most of the times. However, they may spread out and cause systemic infections even if there is a rapid death of cells near the inoculation site. Ringspots with several concentric rings are probably the best documentary of the struggle between plants and viruses.

1. *Chlorotic lesions*: Lesions are discolored or pale. Cells in the lesion are apparently alive. How viruses are restricted in this type of lesions is not very clear.
2. *Necrotic lesions*: Lesions consist of dead cells. This is the result of typical hypersensitive reaction. However, viruses do occasionally escape the local lesion and lead to systemic infection.
3. *Ringspot*: Ringspot is a ring-like lesion. Spectacular ringspots with several concentric rings are often observed.

Systemic symptoms

Systemic symptoms refer to symptoms that develop throughout the plants. They are the results of systemic invasion of the plant host by a virus. When viruses reach the vascular tissues, viruses are transported throughout the plants and causes systemic infection. Depending on the appearance of the symptoms they can be classified as the following:

1. *Mosaic patterns*: Mosaic patterns are more common symptoms in virus infection. Mosaic simply means an alternating pattern of tissues with two different colors. The colors can be light green, dark green, yellow, golden, or even white. If the borders between areas are sharp, then the symptom is referred as mosaic, if the borders are defused, it is called mottle. There are no definite distinctions between the two. Other types of related symptoms are vein clearing or vein banding. The mosaic type symptoms can occurs on leaves, fruit or flowers. On monocotyledon plants, the symptoms are more likely to be of stripes or streaks. The extreme mosaic type of symptoms is yellowing of most of the infected leaves, as in bean golden mosaic virus.

   Dark green islands (DGIs) are a common symptom of plants systemically infected with a mosaic virus. DGIs are clusters of green leaf cells that are free of virus but surrounded by yellow, virus-infected tissue. First, transcripts of a transgene derived from the coat protein of Tamarillo mosaic potyvirus (TaMV) were reduced in DGIs relative to adjacent yellow tissues when the plants were infected with TaMV. Second, nontransgenic plants coinfected with TaMV and a heterologous virus vector carrying TaMV sequences showed reduced titers of the vector in DGIs compared with surrounding tissues. DGIs also were compared with recovered tissue at the top of transgenic plants because recovery has been shown previously to involve PTGS. Cytological analysis of the cells at the junction between recovered and infected tissue was undertaken. The interface between recovered and infected cells had very similar features to those surrounding DGIs. They conclude that DGIs and recovery are related phenomena, differing in their ability to amplify or transport the silencing signal.

2. *Yellowing*: Infection by some viruses may induce yellowing throughout the infected plants. Viruses that cause this type of symptoms are usually phloem limited. Infection of the phloem tissue may disrupt transportation of nutrient and water throughout vascular tissue, resulting general yellowing.
3. *Dwarfing*: Stunting and dwarfing of the infected plants are another common symptom of virus infections. The degree of dwarfing varies. As a general rule, the size of the infected plants is almost always smaller than a healthy one. Interference with plant growth hormone levels is regarded as the most important factor for dwarfing. Other factors include reduced photosynthesis and reduced translocation of nutrients.
4. *Necrosis*: Occasionally, an entire plant infected by a virus may become necrotic. Symptoms usually starts along the veins and whole leaves or plants may die eventually. This type of symptoms is usually caused by a delayed hypersensitivity reaction. The infecting virus has moved out of the entry sites and systemically invaded the entire leaf or plant before the hypersensitivity reaction kicks in.
5. *Malformation*: Change the normal shape of a plant or parts of a plant. Symptoms may include stunting for the whole plant, leaf malformation, twisting, curling, shoestring like leaves, enations on leaves, Gall or tumors formation, rolled either upward or downward.

Interference in Host Metabolisms

Viruses are intercellular parasites with a limited genome, and therefore dependent on the host cell for multiplication and proliferation. Viruses exert negative effects on the hosts by interfering with the host metabolism. Symptoms caused by viral infections are generally considered as results of the following: host defense response to viral infection (chlorotic or necrotic lesions), interference with hormone metabolisms (malformation and dwarfing) and nutrient transport (yellowing and dwarfing), reduced photosynthesis (mosaic, yellowing, and dwarfing), or a combination of the above. However, multiplication of viruses within host cells alone appears to not directly cause diseases. Some viruses may multiply to a high level in the host plants, and yet cause no obvious symptoms..

Hypersensitive reactions

Plants have developed several mechanisms to defend themselves against pathogens. One type of reaction involves rapid cell death around the entry site of a pathogen. This hypersensitive reaction (HR) results in visual symptoms as necrotic lesions in tissues localized around the site of the entry or in entire plants if the hypersensitive reaction is delayed until viruses have spread throughout the plant.

Specific recognition between viral genes and host genes is usually involved in this type of reaction. TMV coat protein gene and the N' gene controlled HR in tobacco is a good example. Severe strains of TMV can systemically infect tobacco plants with N' gene while mild strains induce local lesion (HR) on plants with N' gene. Amino acid substitutions in the coat protein gene were observed in chemically induced mutants of the severe strains of TMV that induce HR phenotypes and in the mutants of mild strains that systemically infect N. sylvestris tobacco. Introduction of a specific point mutation in to TMV U1 strain (HR on N' plants) cause systemic infection of the engineered mutants on plants with N' gene.

Effect on chloroplasts and photosynthesis

Chloroplasts are important factories in green plants. They generate, through photosynthesis, energy and carbohydrate materials essential for plant growth and development. Some viruses, such as turnip yellow mosaic virus, induce obvious clumping and fragmentation of the chloroplasts, and formation of large vesicles.

Viral capsid proteins of TMV and cucumber mosaic virus have been reported to enter into chloroplasts. Altered chloroplasts may be responsible for chlorosis or chlorotic type of symptoms and retarded growth of the infected plants. However, chlorosis induced by cucumber mosaic virus is not accompanied by obvious chloroplast damage. The alteration of cell organelles in wheat leaf cells systemically infected with wheat streak mosaic virus was quantitatively investigated at both light- and electron- microscopy levels. Nuclei significantly increased in size at an early stage of infection and contained dispersed heterochromatin. The nuclear envelope was frequently invaginated and had formed membrane-bound vesicles containing membranes, ribosomes, cylindrical inclusions, virus particles, and, occasionally, fibrils and mitochondria. Chloroplasts in infected tissue were smaller, frequently had extrusions, and contained far less starch than chloroplasts in healthy tissue. Many chloroplasts formed double membrane-bound invaginations in the envelope membrane. Virus particles and mitochondria were found within the invaginations. The viral infection also caused the chloroplast envelope's inner membrane to proliferate and fold repeatedly, dividing the stroma into numerous single membrane-bound fragments. Double membrane-bound elongated tubular structures accumulated in large quantities within the cytoplasm of infected cells.

Disruption of hormone balance

Several hormones are very important in regulating plant development such as cell divisions and cell elongation. Disturbance in the hormone balance results in malformation in plant parts, such as tumor formation, stunted growth, and leaf curling.

Nutrient uptake and translocation

Several phloem-limited viruses induce general yellowing of the entire plants, yet these viruses are not distributed throughout the leaf mesophyll and parenchyma cells. Viral interference of the phloem function such as nutrient uptakes and movement of nutrient from one part of plants to another probably affect this symptom type.

Interaction of Viral Proteins with Translational Apparatus

Several viral proteins are capable of interacting with proteins from the translation apparatus such as initiation factors eIF4E (Leonard et al., 2000; Schaad et al., 2000) and poly A binding protein (Wang et al., 2000). These interactions have the potential of interfering with host protein translation system, thus altering the gene expression pattern. Leonard et al. (2000) reported the interaction between the viral protein linked to the genome (VPg) of turnip mosaic potyvirus (TuMV) and the translation eukaryotic initiation factor eIF(iso)4E of Arabidopsis thaliana and Triticum aestivum (wheat). eIF(iso)4E binds the cap structure m7GpppN of mRNAs and has an important role in the regulation in the initiation of translation. The interaction domain on VPg was mapped to a stretch of 35 amino acids, and substitution of an aspartic acid residue found within this region completely abolished the interaction. The cap analogue m(7)GTP, but not GTP, inhibited VPg-eIF(iso)4E complex formation, suggesting that VPg and cellular mRNAs compete for eIF(iso)4E binding. The biological significance of this interaction was investigated. Brassica perviridis plants were infected with a TuMV infectious cDNA (p35Tunos) and p35TuD77N, a mutant which contained the aspartic acid substitution in the VPg domain that abolished the interaction with eIF(iso) IE. After 20 days, plants bombarded with p35Tunos showed viral symptoms, while plants bombarded with p35TuD77N remained symptomless. These results suggest that VPg-eIF(iso)4E interaction is a critical element for virus production.

Similar interaction between NIa protein and eIF4E was also reported. NIa of potyviruses has been shown to confer host genotype-specific movement functions in plants. Specifically, NIa from tobacco etch virus (TEV)-Oxnard, but not from most other strains, confers the ability to move long distances in Nicotiana tabacum cultivar "V-20." To identify cellular proteins that interact with NIa in a host- or strain-specific manner, Schaad et al. (2000) searched a tomato cDNA library using a yeast two-hybrid system. Ten proteins that interacted with NIa were recovered, with translation initiation factor eIF4E being by far the most common protein identified. Interaction of eIF4E with NIa was shown to be TEV strain-specific. eIF4E from both tomato and tobacco interacted well with NIa from the HAT strain, but not from the Oxnard strain.

Using similar yeast two hybrid searching, Wang et al. (2000) identified host plant proteins associating with the RdRp of zucchini yellow mosaic potyvirus (ZYMV). Several cDNA clones representing a single copy of a poly-(A) binding protein (PABP) gene were isolated from a cucumber (Cucumis sativus L.) leaf cDNA library. Deletion analysis indicated that the C-terminus of the PABP is necessary and sufficient for interaction with the RdRp. This is the first report of an interaction between a viral replicase and PABP and may implicate a role for host PABP in the potyviral infection process.

Change of Host Gene Expression through Gene Silencing and DNA Methylation

RNA silencing (post-transcriptional gene silencing) and DNA methylation have been suggested as host defenses against invasion of foreign nucleic acids including viruses. While these mechanisms work well in silencing the invading nucleic acids, they might also silence host genes with homologies to the invading nucleic acids. Both RNA-directed DNA methylation (Wassenegger et al., 1994; Pelissier and Wassenegger, 2000) and DNA-directed DNA methylation (Al-Kaff et al., 1998) have been reported. The following example describes an RNA-directed DNA methylation. Wassenegger et al (1994) introduced one monomeric and three oligomeric potato spindle tuber viroid (PSTVd) cDNA units into the tobacco genome. Southern analysis of the transgenic plants revealed that only their PSTVd-specific sequences become fully methylated. Viroid cDNA methylation could only be observed after autonomous viroid RNA-RNA replication had taken place in these plants. These findings demonstrate that a mechanism of de novo methylation of genes might exist that can be induced and targeted in a sequence-specific manner by their own mRNA. A recent study by Pelissier and Wassenegger (2000) showed that DNA with as little as 30-nucleotide sequence complementarity to PSTVd can be methylated upon infection by the viroid. Although this sequence specific methylation is only limited to the viroid DNA in the genome in this study, it suggests another mechanism that viroids can cause diseases. Production of viral or viroid RNA (especially at a high concentration) may turn off host gene(s) with localized homologous sequences.

Inhibition and Stimulation of Host Gene Expressions

Viral infections have been shown to both inhibit and stimulate host gene expressions. The best studied system in this respect is the infection of pea seed-borne mosaic virus (PSbMV) in pea cotyledon tissues. Wang and Maule (1995) reported that replication of PSbMV in the tissues could transiently inhibit host gene expression. Replication of PSbMV was restricted largely to a zone of cell close to the infection front. In that replication zone, expression of normal host mRNA was shut down completely. However, host mRNAs were expressed in virus-infected tissues outside of the replication zone. By examining the front of virus invasion in immature pea embryos, two heat-inducible genes encoding HSP70 and polyubiquitin were induced coordinately with the onset of virus replication. At the same time, two genes encoding lipoxygenase and heat shock cognate protein were down regulated. The down-regulation was part of a general suppression of host gene expression that may be achieved through the degradation of host transcripts. Their results indicate that, in contrast to some animal virus infections, there is not a general inhibition of translation of host mRNAs following PSbMV infection. This selective control of host gene expression was observed in all cell types of the embryo and identifies mechanisms of cellular disruption that could act as triggers for symptom expression. Similar host gene induction and down regulation were also observed with Pea early browning tobravirus, White clover mosaic potexvirus, and Beet curly top geminivirus.

Interference with Cell Cycle

The geminivirus contains single-stranded DNA genome and replicates in differentiated plant cells. These cells have left the cell division cycle and contain no detectable levels of DNA replication enzymes or proliferating cell nuclear antigen (PCNA) which associates with DNA polymerase delta and promote its processivity. Nagar et al. (1995) showed that tomato golden mosaic virus (TGMV) induces the accumulation of PCNA in mature cells of Nicotiana benthamiana. Later on, Egelkrout et al. (2001) determined that PCNA protein accumulation reflected transcriptional activation of the host gene. The PCNA transcript of approximately 1200 nucleotides was detected in young leaves, but not in mature leaves of healthy plants. However, the same transcript was found in mature leaves of infected plants. Analysis of the PCNA promoter showed that it was induced during TGMV infection. Developmental studies established a strong relationship between symptom severity, viral DNA accumulation, PCNA promoter activity, and endogenous PCNA mRNA levels. Together, their results demonstrate that geminivirus infection induces the accumulation of a host replication factor by activating transcription of its gene in mature tissues, most likely by overcoming host-mediated repression.

Meanwhile Xie et al. (1995) revealed how wheat dwarf geminivirus AL1 protein might ractivate the host DNA replication machinery by interacting with the retinoblastoma protein (pRb), a cell cycle control protein. Unlike other DNA virus proteins, AL1 does not contain the pRb binding consensus, LXCXE, and interacts with plant pRb homo logues (pRBR) through a novel amino acid sequence. Kong et al. (2000) mapped the pRBR binding domain of AL1 between amino acids 101 and 180 and identified two mutants that are differentially impacted for AL1-pRBR interactions. Plants infected with the E-N140 mutant, which is wild-type for pRBR binding, developed wild-type symptoms and accumulated viral DNA and AL1 protein in epidermal, mesophyll and vascular cells of mature leaves. Plants inoculated with the KEE146 mutant, which retains 16% pRBR binding activity, only developed chlorosis along the veins, and viral DNA, AL1 protein and the host DNA synthesis factor, proliferating cell nuclear antigen, were localized to vascular tissue. Their results established the importance of AL1-pRBR interactions during geminivirus infection of plants.

Symptoms Induced by Transgenic Expression of Single Viral Genes

With the technological improvement in plant transformation, more and more viral genes are being inserted into plant genomes for one reason or another. Most of viral genes did not confer any phenotypical changes in the transgenic plants. However, some did results in symptoms similar to those of viral infections. Transgenic Nicotiana benthamiana plants expressing squash leaf curl virus (SqLCV) BL1 movement protein gene were abnormal both in their growth properties and phenotypic appearance (Pascal et al., 1993). Leaves from the transgenic plants were mosaic and curled under, symptoms typical of SqLCV infection in this host. On the other hand, transgenic plants expressing movement protein BR1 or a truncated BL1 were phenotypically indistinguishable from wild-type N. benthamiana. BL1 was localized to the cell wall and plasma membrane fractions, whereas BR1 was predominantly in the microsomal membrane fraction. These findings demonstrate that expression of BL1 in transgenic plants is sufficient to produce viral disease symptoms. This observation was later confirmed with tomato mottling geminivirus (TMoV). Tobacco plants transformed with the TMoV movement protein (pathogenicity) gene (BL1) had phenotypes ranging from plants with severe stunting and leaf mottling (resembling geminivirus symptoms) to plants with no visible symptoms (Duan et al., 1997). The sequence data for the BC1 transgene from the transgenic plants with the different phenotypes indicated an association of spontaneously mutated forms of the BC1 gene in the transformed tobacco with symptomatic phenotype variations. Both of the above examples involve viral movement proteins, which are known to interact and increase the size exclusion limit of plasmodesmata. Perhaps the increased size exclusion limit interferes with the movement of signal molecules or even large RNA and protein molecules between plants cells, resulting altered metabolisms. Virus-like symptoms were also reported in transgenic plants expressing the open reading frame 0 (ORF0) of luteoviruses. Van der Wilk et al. (1998) constructed transgenic plants expressing the potato leafroll luteovirus (PLRV) ORF0 p28 protein. ORF0 expressing transgenic potato plants displayed an altered phenotype resembling virus-infected plants. A positive correlation was observed between levels of accumulation of the transgenic transcripts and severity of the phenotypic aberrations observed. In contrast, potato plants transformed with a modified, untranslatable ORF0 sequence were phenotypically indistinguishable from wild-type control plants. Southern blot analysis showed that the transgenic plants that accumulated low levels of ORF0 transcripts detectable only by reverse transcription-polymerase chain reaction, contained methylated ORF0 DNA sequences, indicating down-regulation of the transgene provoked by the putatively unfavorable effects p28 causes in the plant cell.

Viral Recombination in Virus Resistant Transgenic (VRT) Plants

Plant viruses contain only a few genes: a replicase, coat protein, movement protein, perhaps a few others. The genome of the virus is comprised of RNA. In order to construct chimeric genes that direct the expression of individual viral proteins in transgenic plants, the RNA first had to be converted into a DNA copy. An RNA sequence can be copied into a DNA molecule, maintaining the correct sequence of bases, by reverse transcriptase. This is the first step towards making the chimeric genes to express viral proteins, as outlined below.

1. Copy the viral RNA into DNA using reverse transcriptase, make into a double stranded DNA.
2. Clone this DNA into a pslamid and sequence the viral genome to identify the genes in the virus.

3. Construct a chimeric gene consisting of the open reading frame for the coat protein and a strong promoter, such as the 35S promoter, to give high-level expression of this transgene.
4. Produce transgenic plants containing this transgene.

This was first tested with transgenic tobacco plants engineered to express the coat protein of tobacco mosaic virus (TMV). When transgenic tobacco plants that express the coat protein of TMV were exposed to the virus, they were found to be resistant to infection by the virus. The inoculated transgenic plants did not develop symptoms, while the control non-transformed plants did become infected. However, the transgenic plants expressing the TMV coat protein were not completely immune; they would become infected if inoculated with a high dose of the virus, but it took longer for symptoms to develop and the symptoms were less severe.

This protection works against the virus from which the coat protein was isolated, as well as against closely related viruses. Having been shown to work providing protection against TMV in tobacco, the same strategy (expression of the virus coat protein in transgenic plants) was then applied to other plant viruses. This approach has now been shown to work providing resistance against a large number of different plant viruses. This is called coat protein-mediated resistance.

In this is not exactly clear, but it appears to block some early stage in the infection process, maybe the uncoating of the virus when it first enters the plant cell. If transgenic plants are infected expressing coat protein with normal viruses, the plants are protected, but if the plants are infected with the RNA of the virus, the plants will develop symptoms. So it appears that expressing the coat protein in the plant prevents the virus from uncoating, expressing the viral genome and causing disease symptoms.

This method has been used successfully in greenhouse and field tests to obtain resistance to many plant viruses. While the first, and the majority, of experiments and trials have been conducted with model systems such as tobacco, the strategy appears to be equally successful with crop plants. It appears to be a technique, which can be applied very widely to obtain resistance. Another vegetable crop that has been developed with resistance to a virus using coat protein mediated resistance is potato. These are being developed by Monsanto with resistance to potato leafroll virus and potato virus Y. Perhaps the best example to date of this strategy is in trying to develop papaya trees with resistance to papaya ringspot virus (PRSV). This is essentially a lethal diseasefor papaya and prevents commercial production of these tropical fruits. There was a thriving papaya industry in Hawaii, which had to move from one region of the island to another because of the spread of PRSV. Over the last ten years researchers at a number of institutions have been developing transgenic papaya that express the coat protein of PRSV. Clone a cDNA for the PRSV coat protein gene. Construct a chimeric gene where the open reading frame of this coat protein is placed under the control of a high level promoter. Transfer this chimeric gene into papaya by transformation. Regenerate transgenic plants and evaluate for resistance. This has led to the release of a transgewic papaya with resistance to PRSV, which is now being planted in Hawaii. While coat protein-mediated resistance has proven to be quite effective, other methods have also been tested to develop plants with resistance to viruses. When the replicase of the virus is expressed in transgenic plants, this also provides resistance to the virus from which the replicase was cloned. Expression of coat protein and replicase are the best examples of what is called pathogen-derived resistance, where part of the virus is used to make plants resistant to that pathogen.

Other strategies that have been used include:

1. Using antisense RNA against the viral genome.
2. Expressing an enzyme from mammals that modifies the viral RNA, inhibiting its replication
3. Expressing ribozymes, short RNA molecules that are able to bind to specific RNAs in the cell, in this case the viral RNA, and break it into smaller pieces.

All of these have been shown to work to some degree, but none have been developed into a commercial application at this time. The final method for making plants resistant to viruses targets the movement of viruses from one plant cell to another. This intercellular movement of viruses requires that the plasmodesmata, the pores between cells, be enlarged. Many viruses produce amovement protein that is responsible for this modification of plasmodesmata. From detailed studies on viruses like TMV, it was shown that some strains of virus were unable to move from cell to cell and produced very limited disease symptoms. The reason these viruses could not move was because they have a defective movement protein, one that prevented the plasmodesmata from enlarging. When this defective movement protein was expressed in transgenic plants, it was found to provide resistance against viruses, not by preventing the infection of plant cells but by preventing the spread of virus from cell to cell.

The risks associated with the introduction of transgenic plants carrying virus-derived sequences include heterologous encapsidation, synergism and recombination. Heterologous encapsidation ('transencapsidation') describes the complete or partial encapsidation of the genome of one virus with the coat protein (CP) of another virus. Synergism describes a phenomenon in which one virus may complement the effects of other viruses resulting in more severe disease (Maki-Valkama and Valkonen 1999). Virus recombination describes the exchange of genetic material between two co-infecting viruses or between a single infecting virus and a virus-derived transgene with the potential for generating a chimeric virus with a novel pathology. Of these three, only recombination has the potential to persist in the environment in the absence of the transgenic plant.

The intention in writing this review was to provide the reader with an informative digest of our present understanding of the nature and significance of virus recombination in transgenic plants containing virus-derived sequences. To this end, the review is set-out as follows:

1. Evolutionary significance of recombination in plant viruses.
2. Mechanisms of pathogen-derived resistance (PDR) in transgenic plants.
3. Evidence of recombination among plant viruses and in plants.
4. Risks associated with recombination and transgenic plants.
5. References cited.

The review focuses almost entirely on RNA recombination. This is because the vast majority of plant viruses and virus-derived sequences (promoters and transgenes) used in transgenic plants have an RNA origin. While there is presently little published information on other types of genetic exchange involving plant viruses the following generalities can be stated:

1. Recombination among RNA plant viruses appears to be relatively frequent.

2. Recombination among DNA plant viruses appears to be relatively frequent.
3. Recombination between RNA and DNA plant viruses appears to be relatively infrequent.
4. Recombination between RNA or DNA viruses and host plant DNAs appears to be relatively infrequent.

Two types of transgenic plant are relevant to discussions on recombination (i) GM plants in which virus-derived promoters are used to control the expression of non-viral transgenes, and (ii) GM plants in which virus-derived transgenes are used to confer disease resistance in virus resistant transgenic plants (VRTPs). Virus-derived promoters are used in a wide range of plants that are not necessarily hosts of the virus (or its relatives) from which the promoter sequence was originally sourced. Since the key constraint on genetic exchange is the degree of homology (relatedness) shared by donor and recipient sequences (next section) recombination involving virus promoters will only be an issue for relatively few of the plant species in which the promoter is deployed.

Evolutionary Significance of Recombination in Plant RNA Viruses

Two distinct types of genetic exchange operate in RNA viruses. The first, re-assortment, occurs only in segmented viruses and involves the exchange of one or more discrete RNA molecules that comprise the multipartite virus genome. The second type of genetic exchange involves the introduction of a donor nucleotide sequence into a single contiguous acceptor molecule to form a novel recombinant RNA molecule and can occur in segmented as well as unsegmented viruses.

Nomenclature

Based on the similarity of the parental RNA molecules, two types of RNA recombination have been defined: homologous recombination and non-homologous recombination. Homologous recombination (HR) occurs between two similar or identical RNAs involving precisely matched (precise HR) or divergent (aberrant HR) RNA sequences. Non-homologus recombination (NHR) involves the exchange between identical viruses in which strict alignment is not maintained or involves exchange of unrelated sequences. This nomenclature is not entirely satisfactory since the term 'homologous' implies common ancestry.

The recombination mechanism

Two mechanistic models have been proposed to account for RNA recombination:

1. A breakage-rejoining model which supposes the cleavage at two distinct sites of the same or two different precursor RNAs followed by ligation of the two cleaved molecules into one recombinant RNA
2. An RNA dependent RNA polymerase (RdRp) mediated copy-choice model, also known as 'template switching', in which the viral RdRp complex switches from one RNA template to another during replication. Where replication of the new strand continues precisely where it left the first template RNA the recombinant virus is homologous and non-homologous when it does not.

The only experimental evidence in support of the breakage-rejoining model for RNA recombination has come from Chetverin et. al. (1997) from in vitro studies of the Q? bacteriophage in which only non-homologous recombinants were observed. Nevertheless, almost all the other evidence supports the RdRp-mediated copy-choice model (e.g. Kirkegaard and Baltimore 1986; Nagy et. al. 1995; Nagy and Simon 1997; Figlerowicz et. al. 1997 and 1998) although it should also be noted that the template-switching mechanism has yet to be formally demonstrated.

Since template switching occurs during RNA synthesis several physical requirements need to be met for successful cross-over to occur. The first of these is that the RNA polymerase, which is responsible for catalysing the synthesis of the new RNA molecule, must first pause and then dissociate from the template viral RNA (or DNA) molecule. The second condition is that an alternative template must be sufficiently close for the dissociated polymerase to bind to. The third condition is that the new template has a polymerase binding site to facilitate cross-over and continued synthesis on the new RNA molecule (Lai 1992).

Evidence indicates that these conditions are generally met. Transcriptional pausing has been observed in a range of RNA viruses and RNA polymerase binds only weakly to RNA, which increases its potential for dissociation. Also, in most RNA viruses, RNA synthesis occurs in membrane bound cellular compartments and this may promote the local concentration of viral RNAs making template switching by polymerase more likely. The tendency for some regions of an RNA molecule to form double stranded 'heteroduplexes' may also serve to promote both transcriptional pausing and physical bridging of the template RNAs by the polymerase. The mechanism by which polymerase recognises binding sites on an alternative RNA template probably involves either sequence complementarity (where homologous recombination occurs), or recognition of common RNA secondary structure where non-homologous recombination occurs, or both (Lai 1992; Aaziz and Tepfer 1999).

Evolutionary advantages of recombination

Evidence derived in the main from nucleotide sequence data has been used to compile a large and rapidly growing list of RNA viruses that have evolved as a result of genetic recombination. It is now unequivocal that recombination has played a central role in the evolution of many RNA viruses. Theoretical explanations of the evolution of sexual reproduction tend to emphasize the spread of advantageous genes and the removal of deleterious genes as the main drivers. The second of these derives from 'Mueller's ratchet' theory which predicts the gradual build-up of deleterious alleles in finite asexual populations. RNA viruses represent one of only a few examples that have been used to empirically test this prediction and results have generally indicated agreement i.e. decreased fitness in populations in the absence of reassortment.

Direct experimental evidence supporting a similar role for recombination has not been provided to date, but expectations are that genetic exchange through recombination provides for a similar function to reassortment by fixing advantageous alleles in the population and removing disadvantageous alleles by combining mutation free parts of different genomes. Although it has been proposed that recombination in monopartite RNA viruses and reassortment in segmented RNA viruses represent alternative evolutionary pathways in these two groups. The weight of evidence for plant viruses, as well as other RNA viruses, indicate that these two mechanisms are not necessarily exclusive and in some cases may even operate simultaneously.

The evidence that has been presented in favour of the theory that recombination can eliminate deleterious alleles has come from studies that show that weakly replicative mutant tombusviruses, for example, can be reconstituted through recombination with intact wild-type homologues (White and Morris 1994). Similar evidence has also been generated for the bromoviruses (Rao and Hall 1993). Several landmark studies have also shown that mutant viruses can repair their genomes through genetic recombination with virus transcripts in VRT plants carrying viral derived transgenes (pages 38-42). Recombination between viruses and host transgenes under conditions of strong selection has been shown not only to produce viruses that cause disease symptoms quite distinct from the parental strain (Green and Allison 1994), but also hybrid viruses more pathogenic than the parental strains (Jacquemond et al. 1997). Two important studies have shown that RNA recombination in plant viruses may even provide a telomerase-like function by repair to the 3' ends of satellite RNAs, for example, of turnip crinkle virus (Burgyan and Garcia-Arenal 1998) and cucumber mosaic virus (Simon and Nagy 1996).

In addition, a growing body of information indicates that recombination between viral and host genomes has occurred (as illustrated by sequence comparison) between a luteovirus and a plant chloroplast exon (Mayo and Jolly 1991). Closteroviruses have also contributed supporting evidence with sequence data indicating that this group has evolved in part by acquiring various host protein-coding genes (Dolja et. al. 1994). Evidence for the uptake of viral sequences by plant hosts is also beginning to emerge with a report of the suspected integration of Banana Streak Badnavirus (BSB) into the Musa genome (Harper et. al. 1999). Presumably, the selective advantages to host and pathogen are resistance to or amelioration of disease and in maximising disease transmission, respectively.

Recombination among RNA viruses is probably relatively common but only hybrids for which the benefits of exchange outweigh the costs and for which the hybrid is able to compete with parental strains are likely to persist. As with mutation, genetic exchange is a random process and the vast majority of new combinations will either be non-functional or suffer decreased fitness compared to the parental viruses. As such, the vast majority of recombinants are likely to be rapidly eliminated from the virus gene pool.

Mechanisms of Pathogen-Derived Resistance (PDR) in Transgenic Plants

The incorporation of viral-derived sequences into plants has provided the means of combating economically significant crop viruses for which there are no readily accessible sources of natural resistance. Only two plant genes demonstrating resistance to viruses have been isolated to date. These include the N gene from tobacco (Erikson et. al. 1999) and the Rx gene from potato (Bendahmane et. al. 1995) which encodes a hypersensitive response to tobacco mosaic virus (TMV) and an extreme resistance potato virus X (PVX) respectively. However, the mechanism by which each of these responses is induced is not known and while the N gene has been successfully transferred to tomato, the wider applicability of these transgenes to other crop plants has not been demonstrated.

Strategies for pathogen-derived resistance (PDR) are divided into those that require the production of proteins and those that require only the presence of RNA sequences. Generally speaking, protein-mediated strategies provide incomplete resistance to a wide range of RNA viruses, whereas RNA-mediated resistance provides very high levels of resistance to a narrower range of viruses (Beachy 1997).

Protein Mediated Resistance

Coat protein (CP) mediated resistance

Coat protein mediated resistance (CPMR) can provide either broad or narrow resistance to RNA viruses. The coat protein of TMV for instance is effective against closely related strains, but resistance decreases with diminishing sequence similarity. In contrast, the CP gene of soybean mosaic virus (SMV) which is not capable of infecting tobacco, confers resistance in tobacco to two unrelated potyviruses, potato virus Y (PVY) and tobacco etch virus (TEV). It is not known why some coat proteins afford high levels of protection and why others provide broader or lower levels of resistance, though this issue will almost certainly have direct relevance to the risks of virus recombination where transgenes from several virus strains are required to deliver sufficiently broad resistance.

Replicase - mediated resistance

Genes encoding viral replicase proteins can confer near immunity to infection. All plant viruses have replicase genes and transformation of plants with both defective and non-defective replicase genes can provide very strong resistance - though usually only to the donor virus or very closely related strains. The mechanisms involved in replicase-MR are not known but are likely to involve the suppression of virus replication and gene expression.

While a feature of replicase-MR is that it is generally quite specific, some interesting exceptions have been observed. In one case, transgenic resistance to potato leaf roll virus (PLRV) turned out to be effective against a wide range of PLRV strains. In another example, the TMV replicase gene containing an Agrobacterium sequence insertion was found to confer high level resistance in tobacco plants to a wide range of tobamoviruses (Donson et. al. 1993).

Movement protein - mediated resistance

Movement protein mediated resistance (MPMR) is thought to arise through competition between the dysfunctional introduced MP and the invading virus for binding sites on host plasmodesmata resulting in reduced local and systemic spread of infection in MPMR transgenic plants. In contrast to replicase-MR, the expression of dysfunctional MP genes facilitates resistance to a much wider range of viruses (Malyshenko et. al. 1993). This approach has not seen wide application due to the scale of effort necessary to make and screen banks of mutant MP genes to identify effective candidate mutants.

The expression of functional MPs appears either to have no effect or actually increases susceptibility of plants to virus disease (Ziegler-Graff et. al. 1991). It has been anticipated that improved knowledge of MP structure and function will lead to the development of mutant MPs that can act as highly effective inhibitors of many different types of virus (Beachy 1997).

RNA-mediated resistance

Several PDR strategies involve the expression of genes that do not encode proteins. One of the first was expression of antisense RNA sequences to inhibit virus replication (Hammond and Kamo 1995) although some doubt exists as to whether this effect was actually due to RNA suppression. Other mechanisms may also be responsible including interruption of template selection by viral replicase and/or the formation and subsequent degradation of double-stranded RNA. Antisense mediated suppression is generally very narrow in terms of the range of viruses that it targets.

RNA mediated resistance in transgenic plants has also been shown to operate through competitive inhibition of the invading virus with the transgene or its RNA transcript acting as a decoy for proteins needed for virus replication and movement. This process is known to operate in instances where the transgene specifies a 'defective-interfering' (DI) viral RNA as, for example, in transgenically expressed turnip yellow mosaic virus (TYMV) 3' DI RNA which competes with the invading TYMV genome (Zaccomer et.al. 1993). Satellite RNAs are similar to DI molecules but do not depend on sequence similarity with the donor virus. Transgenic plants encoding DI RNAs (or DNAs) and satellite RNAs have been tested for their capacity to reduce replication and ameliorate disease with some degree of success. Presently, the most well known type of RNA mediated resistance is RNA suppression - a process that describes the post-transcriptional destruction of viral RNA also known as "gene silencing" (Baulcombe 1996). Virus gene silencing was suspected early on in the development of VRT plants from observations that in plants designed to be resistant through CP expression; protection was maintained even in the absence of the transprotein (Lindbo et. al. 1993). Gene silencing in plants derives either from a block on the initiation of viral transcription or from degradation of viral RNA after transcription and is described as transcriptional gene silencing (TGS) and post-transcriptional gene silencing (PTGS) respectively. In either case, the host cell activates a destruction mechanism upon detecting aberrant RNA sequences and both viral and transgene RNA is destroyed.

Suppression of RNA Silencing

Suppression of RNA silencing probably represents an evolutionary adaptation by plant viruses to a generic host antiviral defence strategy since many of the suppressor proteins have previously been shown to be important determinants of virulence. In fact, the ability to suppress PTGS may be crucial for many viruses to be able to infect their plant hosts. It seems that suppression occurs in a general manner, independent of the gene or sequence responsible for inducing the silencing mechanism that, by contrast, can only act against near homologous sequences.

Transgenic plants designed to work by gene silencing are only effective because the PGTS mechanism is activated prior to infection and is therefore able to degrade the target RNA before viral proteins can be translated from it.

Evidence of Recombination among Plant RNA Viruses and Transgenic Plants

Mutation, reassortment and recombination are the drivers for evolution in RNA viruses. Recombination probably plays an additional role in rescuing viruses by repairing mutations in virus genomes that have occurred as a result of proof-reading errors during replication (Carpenter and Simon 1996). The realisation that recombination has a central role in the evolution of many RNA viruses began with experiments on poliovirus by Hirst (1962) and Ledinko (1963). Since that time recombination has been documented for many animal and plant RNA viruses. The rapid accumulation of evidence for recombination is all the more impressive when one realises that prior to these formative studies, recombination was not thought to be a property of RNA genomes at all.

Recombination in plant RNA viruses was first demonstrated for brome mosaic virus (BMV) and there is now an extensive literature documenting recombination among both plant and animal RNA viruses as well as in bacteriophages and double-stranded and negative sense RNA viruses. The genetic engineering revolution, which has brought with it the development of the polymerase chain reaction (PCR), automated nucleotide sequencing techniques and phylogenetic analysis, has transformed the study of virus recombination by providing the means of detecting and characterising recombination events. Using these techniques it has been possible not only to detect rare recombination events and exchanges that have occurred long ago (e.g. Weaver et. al. 1997) but also to identify the precise location of the cross-over point (Olsthoorn and Duin 1996). Two types of data have been used to provide evidence of RNA recombination, one based on phylogenetic analysis and the other based on empirical observation.

Detecting Recombination among RNA Viruses

Methods for analysing putative recombination events are often graphical in nature and exploit the fact that many recombinant sequences are mosaics of virus phylogenies. For example, 'Split Decomposition' analysis (Huson 1998) employs a split graph highlighting transition from a dichotomously branching tree form indicative of sequence relationships to a complicated network indicative of a history of genetic exchange. Other graphical programmes look for discordant nucleotide sequence relationships suggestive of recombination, examples of which include 'boot scanning' (Salminen et. al. 1995), 'PhylPro' (Weiller 1998), 'TOPAL' (McGuire and Wright 1998) and 'DIVERT' (Gao et. al. 1998).

Statistical approaches employ procedures based on maximum-likelihood computation of goodness-of-fit estimates by comparing the frequency distribution of polymorphisms in putative parental and recombinant strains. Some statistical approaches attempt to quantify the amount of recombination rather than document specific recombination events and do this by calculating an 'Index of Association' which measures the degree of linkage between alleles at different loci. A further development on this approach is the 'Homoplasy Test' which compares the number of convergent and parallel base changes (homoplasies) in a maximum parsimony tree to the number expected by chance alone where excessive homoplasy indicates a history of recombination.

There is also an abundance of direct experimental data demonstrating homologous RNA recombination using movement-defective viruses in which functional recombinants have been restored through genetic exchange. Non-homologous recombination (NHR) between plant viruses has been demonstrated in two instances. The first involved co-inoculation of protoplasts and plants with defective cucumber necrosis virus (CNV) and tomato bushy stunt virus (TBSV) RNAs which produced recombinant defective interfering (DI) strains as well as fully infectious derivatives. The second case involved the co-inoculation of tomato aspermy virus (TAV) RNAs 1 and 2 and cucumber mosaic virus (CaMV) RNAs 2 and 3, resulting in a novel cucumovirus encoding a novel replicase comprising TAV helicase and CaMV polymerase subunits (Masuta et. al. 1998).

The phylogenetic evidence as well as the empirical evidence indicates that not only do different RNA viruses not recombine at the same frequency, but also that some strains are highly recombinative while others show no evidence of recombination at all. Phylogenetic analyses of potyvirus strains, for example, indicate numerous recombination events, whereas analyses of cucumovirus genomes have yielded no evidence at all have past recombination events. While these findings may be explained to some extent by differences in the detection limits of the different virus systems, it is also likely to be due, in part at least, to the inherent characteristics of the virus-host interaction. This last point, in particular, is significant for VRT plant risk assessment since almost nothing is known about the role of host factors in viral RNA recombination.

Recombination in VRT Plants

Numerous transgenic crops have been developed for tolerance or complete resistance to a range of economically important virus diseases. Recombination in transgenic plants has been investigated in systems in which movement defective viruses are inoculated into plants expressing the functional homologous movement protein gene, and then tested for the presence of systemic (recombinant) viruses. This phenomenon has been reported in three experimental systems: cauliflower mosaic virus (CaMV), cowpea chlorotic mottle bromovirus (CCMV) and tomato bushy stunt tombusvirus (TBSV).

The first extensively documented case of recombination involving a transgene was in Brassica napus expressing the cauliflower mosaic virus (CaMV) translational transactivator gene ORF VI. Experimental plants were inoculated with a mutant CaMV strain lacking the ORF VI gene and recombination with the transgene restored systemic movement to the inoculated strain. In an important related study a CaMV strain unable to infect solanaceous plants acquired ORF VI from transgenic tobacco, Nicotiana bigelovii, generating recombinants with altered symptoms and extended host range. A strain of cowpea chlorotic mottle bromovirus (CCMV) that had been crippled by deletion of part of the coat protein gene was used to elegantly demonstrate recombination in transgenic N. benthamiana containing the missing CP sequence (Green and Allison 1994, 1996). In this example, mutation markers unequivocally confirmed a single recombination event between the transgene mRNA and the movement-defective CCMV to generate fully restored systemic hybrids containing the functional CP gene. Sequence analysis of the recombinant strains revealed numerous modifications flanking the putative cross-over site, and when tested on a range of host species, four out of seven recombinants displayed novel symptoms, although none was more fit than the wild-type virus (Allison et. al. 1997 and 1999). In these experiments, recombinant viruses were recovered under conditions of strong selection pressure in that the challenging virus was movement defective and had to recombine with the transgene in order to regain function. Searching for recombinant viruses generated under conditions of little or no selection pressure is generally regarded as necessary for realistic risk assessment.

Perceived Risks Associated with Virus Recombination and Transgenic Plants

The key characteristic that distinguishes genetically modified organisms (GMOs) from most other technologies (e.g. chemical pesticides) is that GMOs can replicate and, consequently, once established in the environment a GMO may be able to persist indefinitely and prove difficult and costly to remove if this becomes necessary. Persistence does not just imply temporal continuity but also indicates spatial spread since not only could the entire GMO disperse beyond the boundary of the intended release site but the transgene itself may be able to disperse into the wider environment as a result of out-crossing between the GM plant and compatible wild relatives.

Uptake of the transgene by wild-type organisms provides the potential for unique and unintended combinations of genes and for this reason considerable attention is now being directed at understanding the risks associated with gene introgression. In addition to out-crossing, VRT plants present an additional concern because of the possibility that the viral derived transgenes used to confer disease resistance in GM crops may recombine with the genomes of naturally occurring plant viruses.

Ecological Risk Experiments

There are few published ecological studies assessing the potential for environmental impact from the cultivation of virus recombination in VRT plants. Agriculturally important plant viruses are capable of impacting on natural host plant populations and that impacts are unlikely to be uniformly distributed among host populations. This was almost certainly as a result of differences in the distribution and behaviour of the aphid, weevil and flea vectors responsible for transmitting these virus pathotypes. Since evidence from the controlled trials discounted the possibility that vector species had imposed differential selection for virus resistance among the study populations, the conclusion was that the potential for virus induced impacts on wild B. oleracea populations will be highly dependant on the distribution and behaviour of the vector species involved.

Viral Diseases of Maize

The most important viral diseases of maize are maize dwarf mosaic (MDM) and maize chlorotic dwarf (MCD). Yield losses in fields where the disease is well established may be severe depending on the susceptibility of the corn hybrid being grown. In general, most loss occurs when the plants become infected at the "knee high" stage of development and is minimal on most hybrids if infection occurs after tasseling. Diagnosis of virus infected plants is difficult based on field symptoms. Samples of plants should be tested in the laboratory to confirm the presence of the virus.

Maize dwarf mosaic (Symptoms)

1. Maize dwarf mosaic symptoms vary with the corn hybrid and the stage of development of the plant at the time of infection.
2. Early infection results in chlorotic spots or flecks that elongate in young leaves in the whorl. Flecks merge into chlorotic streaks along the leaves.
3. These streaks form mosaic or mottled patterns and may turn to a general yellowing as the growing season progresses.
4. Later, plants may have blotches or streaks of red that generally appear after periods of cool (60°F) night temperatures.
5. Infected plants are predisposed to root rot and may be barren.

Fig. 5.1. Maize dwarf mosaic.

Disease cycle

Maize dwarf mosaic virus (MDMV) exists in several strains, the most common being strain A, which infects and overwinters primarily on johnsongrass. Strain B does not infect johnsongrass. Besides corn and johnsongrass, the MDM virus may infect over 100 wild and cultivated grasses. More than 20 species of aphids can transmit MDMV. An aphid can acquire the virus within a few minutes of feeding on an infected corn or johnsongrass plant. The aphid then flies or is carried by the wind to other corn plants and inoculates them with the virus. The aphid retains, and is generally able to transmit the virus for 15-30 minutes after acquiring it. The corn leaf aphid and the green peach aphid are common aphid vectors of MDMV.

Maize chlorotic dwarf (Symptoms)

1. Maize chlorotic dwarf symptoms include yellowing of youngest leaves in the whorl and a distinct fine yellow striping, or vein clearing of the smallest veins visible between the larger veins.
2. This chlorotic vein clearing of secondary veins is diagnostic for maize chlorotic dwarf and may be more readily observed on the undersides of infected leaves.
3. The striping may not be seen in later stages of development due to leaf reddening and general yellowing of the plant.
4. Affected plants may also be stunted due to shortening of the upper internodes.

Disease cycle

Maize chlorotic dwarf virus (MCDV) is primarily transmitted by the black-faced leafhopper. The leafhopper must feed on an infected plant for several hours to acquire MCDV and then can transmit it for up to 48 hours.

Control Measures

1. Grow hybrids tolerant or resistant to maize dwarf mosaic. There is good tolerance and resistance to strain A, but only fair tolerance and no resistance to strain B in dent corn. There is no resistance to maize chlorotic dwarf virus and only fair tolerance.
2. Destroy grasses hosts, including volunteer corn in areas where corn is to be planted.
3. Plant early, since early-planted corn will escape damage because aphid populations do not build up until the plants are past the seedling stage.

Viral Diseases of Peanut

Unlike many of the pathogens of peanut that depend on the forces of wind or rain for dissemination, a vector actively disseminates most viruses. Insects such as aphids or thrips are common vectors, but people and equipment may also spread viruses by mechanical damage of plants. Yield losses from virus diseases in peanut depend on the severity of the disease, the percentage of plants infected, and the plant age when infection occurs. Yield loss is generally high if a large number of plants are severely infected early in the growing season. Yield loss is usually minimal when infection occurs late in the season when, most or all of the pods are mature, or by mild viruses, regardless of when plants are infected.

Peanut mottle (Symptoms)

1. The symptoms of PMV infection can vary with cultivar, time of infection and environment. The most common symptom, although it may not be readily noticed, is a mild mottle or mosaic on the youngest leaves of infected plants.
2. The light and dark green areas of affected leaves can best be seen if leaves are held up to light.
3. Margins of leaflets may curl up and depressions in the leaf tissue between the veins may become prominent. Plants are generally only mildly stunted, if at all.
4. As plants mature, the symptom expression generally declines, particularly during hot and dry weather.
5. Pods from infected plants may be reduced in size and have irregular gray to brown patches. The seed coat of affected seed may also be discolored.

Disease cycle

Peanut seed is thought to be the most important primary source of PMV for the beginning of the disease cycle. Commercial seed lots usually have a low frequency of PMV infection (less than 1 percent). However, an infection level of 0.1 percent can result in about two infected seedlings per 100 square yards in a field. Aphids may then transmit the virus from infected to healthy seedlings. At least six species of aphids mechanically transmit PMV. Winter forage legumes such as clovers are also hosts for PMV and may serve as sources for the virus. can efficiently transmit

Fig. 5.2. Mottled leaf caused by infection with Peanut mottle virus.

the virus. Tractor tires crushing infected vines in touch with healthy vines may also.

Control measures

1. Because the virus generally has a minor effect on commercial peanut yields, no specific management practices are recommended for control of PMV.
2. Several breeding lines with resistance have been identified, but this resistance has not been incorporated into any commercial varieties.
3. The use of PMV-free seed is the most feasible approach for control, as this prevents the disease from becoming initially established in the field. PMV-free seed can be produced under rigorous procedures. If PMV-free seed is used, volunteer plants must be completely removed and the field situated so that PMV hosts, such as clovers, southern pea, and navy bean are at least 100 yards away.

Peanut Stunt (Symptoms)

1. Peanut Stunt virus(PSV) causes striking and destructive symptoms on peanut. The most common symptom of PSV infection is severe stunting of the plant.
2. Stunting may affect the entire plant or a portion of the plant, such as a single branch. 3. Infection early in the season generally results in the most severe expression of symptoms.
4. Leaves of infected plants may be curled upward and be malformed. Leaves may also appear yellowed, mottled, or have a dark green banding pattern.

Fig. 5.3. Peanut plant showing stunting leaf distortion, and mottling symptoms caused by Tomato spotted with virus.

5. Pod production is reduced in both size and number. Pods may also be malformed and hulls may split.
6. Seed germination and seedling vigor is reduced when seed from infected plants.

Disease cycle:

PSV is introduced into fields either by infected seed or by aphids, which acquire the virus from another plant host plant, such as white clover. PSV infects at least 115 species of plants. Important economic hosts include peanut, bean, southern pea, and white clover. The virus can be spread mechanically, by aphids, and in seed. Transmission in seed is not thought to be of great significance, because the percentage of transmission from seed is low (less than 0.3 percent) and plant viability from infected seed is low. Aphids, however, play an important role in spread of the virus. Three species of aphids are known to transmit PSV. Aphids may acquire the virus from infected hosts such as white clover or weeds such as crownvetch. The virus may then be carried to nearby peanut fields.

Control measures

1. Peanuts should be planted in fields free of PSV and away from known reservoirs of PSV, such as white clover.
2. The use of peanut seed free of PSV is also recommended.
3. Infected plants should be rogued from seed fields.
4. Cultivars with resistance to PSV have not been developed, although some breeding lines with resistance have been identified.

Tomato spotted wilt (Symptoms)

1. Symptoms of TSWV in peanut can be quite variable. A common symptom of the virus in peanut is the development of yellow to pale green ringspots of various sizes and shapes on infected leaves.
2. These are best seen when infected leaves are held up to light.
3. Stunting also occurs and can be quite severe if plants are infected early in their growth.
4. Infected plants may produce distorted, mottled, and smaller than normal leaflets.
5. Plants infected within 45 days of planting produce few, if any, marketable pods. Yield losses decrease when as progressively

older plants are infected. Late-season infections usually have little impact on yield.

6. Death of terminal buds may occur and plants may be killed in extreme cases.
7. Dead plants are usually found in fields with a high incidence of plants that were infected early.
8. Seed produced on infected plants may be reduced in size, malformed, and have discolored seed coats.

Fig. 5.4. Peanut leaf with ringspots caused by infection with tomato spotted wilt virus.

Disease cycle

TSWV is not transmitted to plants through infected peanut seed. TSWV infects more than 200 species of plants. These plants include economic crops such as tomato and pepper as well as weed species. Winter annuals such as chickweed, shepardspurse, swinecress, and sowthistle are known to be TSWV hosts. Weeds may serve as a reservoir of the virus in areas affected by tomato spotted wilt. The virus is spread by at least five species of thrips. Thrips become infective only when they acquire the virus while feeding on infected plants as larvae. Larvae can transmit the virus before they pupate, but more commonly adults transmit the virus. Adults remain infective throughout their life. Thrips are generally a widespread insect pest of peanut. Therefore, the potential exists for TSWV to be an important disease in peanut each season.

Control measures

1. Thrips control with granular insecticides is usually recommended where losses from TSWV have occurred in peanut. The success of thrips control in managing TSWV has been variable. This is perhaps because these insecticides are systemic and

infective thrips are able to feed for short periods and infect plants before the ingested insecticide acts to kill them.

2. Maintenance of weed control in areas adjacent to peanut fields may help reduce potential reservoirs of the virus.

Viral Diseases of Bean

Bean common mosaic virus (BCMV): Symptoms

1. Systemically infected plants, especially those from infected seeds, have leaves with green mosaic patterns and distortions (curling, strapping, or puckering of tissues along leaf veins).
2. Plants may be stunted and have only a few pods which mature later than uninfected pods.
3. Vascular tissue can become necrotic, producing dark streaks on petioles and stems.
4. At high temperatures (above 78 F), cultivars with the hypersensitive resistance gene (I gene) respond to necrosis-inducing strains of BCMV with a systemic necrosis called black root. Plants with black root die.
5. Nonsystemic infections can appear as ring-like lesions on foliage.
6. Appearance and severity of symptoms depends on strain of the virus, variety, time of infection, and environmental conditions.

Disease cycle

BCMV is a seedborne virus transmitted by aphids and mechanically by plant sap. Many types of beans, alfalfa, and common clover are hosts.

Control measures

1. More than 15 strains of BCMV are known and breeders have incorporated resistance to the more important strains in many commercial varieties.
2. Certified seed programs are restrictive for BCMV contamination.
3. Controlling large populations of aphids can reduce spread.
4. The primary control is selection of high quality, virus-tested seed of genetically resistant varieties.

Bean yellow mosaic virus (BYMV): Symptoms

1. It causes plant stunting and leaves with contrasting areas of dark green and yellowed tissue.
2. Bright yellow spots can be obvious on older plants.

Disease cycle

This virus is readily transmitted by aphids and mechanically by plant sap, but it is not seedborne. The virus also attacks wild hosts such as clover and sweet clover. Some viral and fungal diseases of broad-acre crops can only survive from the end of one winter growing season to the start of the next on green plant material - the so-called 'green bridge'. The aphid vectors of many crop viruses also require this 'green bridge' to survive between crop growing seasons.

Control measures

1. There is no specific management practices are recommended for control of BYMV.
2. Genes for resistance have been identified.

Viral diseases of clovers (Symptoms)

1. Virus symptoms on clovers vary somewhat, depending on host species and the virus involved.
2. In general, viral infected plants have mild to severe leaf mottling.
3. Narrow, pale to yellow discolored areas may be found along the veins, or large, light green to yellowish blotches may occur between the leaf veins.
4. In some cases, leavescurl, or are puckered or ruffled.
5. Severely affected plants may be dwarfed or weakened and unable to withstand prolonged drought or severe winters.
6. The weakening caused by viral infection may predispose plants to attack by other disease causing agents.
7. Symptoms of most clover virus diseases are conspicuous during cooler periods, disappearing temporarily during hot weather.

   Some examples of viral diseases of commonly grown clovers:

(a) *Red clover*. Red clover vein mosaic is one of the most prevalent and widely distributed viruses of this host. In the field, red clover also is a natural host for yellow bean mosaic and potato yellow dwarf viruses. The viruses of alfalfa mosaic, pea common mosaic, and white clover mosaic also can infect red clover.

(b) *Alsike clover*. Virus mosaic of alsike clover is widely distributed. In addition, alsike clover is known to harbor viruses of annual legumes, principally those of peas.

(c) *White clover* (including Ladino clover). In the field, white and Ladino clovers display at least two kinds of viral disease symptoms. Some plants develop vein clearing with mild mottling while others develop bright yellow patches or streaks between the veins of leaves. The vein clearing and mottling symptoms are commonly caused by white clover mosaic, which consists of a mixture of two viruses, pea mottle and pea wilt. Yellow blotches between the veins, particularly noticeable in Ladino clover, are caused by a strain of alfalfa mosaic virus. A similar strain of virus from alfalfa and Ladino clover causes a severe tuber necrosis in potatoes.

(d) *Sweet clover*. Many viruses can infect sweet clover in the field. For example, the following have been recovered from infected plants: alfalfa mosaic, yellow bean mosaic, several pea viruses, red clover vein mosaic, white clover mosaic, tobacco streak virus, and the Colorado rednode virus of bean.

Disease cycle

The viruses that attack clovers can also infect many related and unrelated hosts. In turn, some viruses from hosts such as peas, beans, potatoes, alfalfa, and certain weeds are readily transmitted to clovers. Thus, a virus-infected crop in one field or plants along a fencerow may act as the reservoir for virus infection of an entirely different crop in a neighboring field. The spread and severity of infection depend on insect populations carrying viruses from plant to plant, the relative susceptibility of the host to the viruses present, and the age of the plants when infected.

Control measures

1. To date, very little has been done to control viral diseases of clover.
2. Where the crop is being grown for seed, insecticides may be applied to reduce the number of insect carriers; clover viruses are rarely seedborne. When possible, clover fields should not be planted adjacent to other leguminous crops such as peas or beans.
3. The ultimate solution is the development of varieties of clovers resistant to the most prevalent and damaging viral diseases.

Viral Diseases of Soybean

Soybean mosaic (Symptoms)

1. Symptoms of SMV vary depending on the soybean cultivar, the age of the soybeans, the virus strain, and the temperature.

2. Symptoms are most noticeable under cool temperatures of 18 to 24°C. When temperatures rise above 30°C, leaf symptoms may be masked.
3. The youngest and most rapidly growing leaves show the most severe symptoms. The leaves of SMV infected plants are distorted and narrower than normal, and develop dark green swellings along the veins.
4. Infected leaflets are puckered and curl down at the margin.
5. Plants infected early in the season are stunted, with shortened petioles and internodes.
6. Diseased seed pods are often smaller, flattened, less pubescence, and curved more acutely than pods of healthy plants.
7. Infected seed are mottled brown or black, usually smaller than seeds from healthy plants, and germination may be reduced.

Fig. 5.5. Soybean mosaic.

Disease cycle

Soybean mosaic is caused by soybean mosaic virus (SMV) and is the most widely distributed virus diseases of soybeans. It is spread by planting diseased seed and by at least 31 species of aphids

Bud blight of Soybean (Symptoms)

1. Infected plants are stunted and the pith of stems and branches show a brown discoloration, first near the nodes, then throughout the stem.
2. Leaves on infected plants are smaller, wrinkled, and have a bronze discoloration.

3. Buds become brown, necrotic, and brittle; hence the name bud blight.
4. The most strirking symptom is the curving of the terminal bud to form a crook.
5. Diseased pods are often aborted or contain no seed. In the field, plants that are infected remain green after healthy plants have matured.

Fig. 5.6. Bud blight of soybean caused by the tobacco ringspot virus (TRSV).

Disease cycle

Bud blight, caused by the tobacco ringspot virus (TRSV), can be a serious disease of soybeans. Yields may be reduced 25-100% depending on the time of infection. Planting infected seed spreads the virus, but the amount of infected seed produced is usually extremely low.

Bean pod mottle (Symptoms)

1. Plants infected with both soybean mosaic and pod mottle are stunted, have distorted foliage, misshapen fruit, and necrotic tissue.
2. Seed from plants infected with pod mottle are smaller than normal.

Disease cycle

This disease, caused by the bean pod mottle virus (BPMV), often occurs in combination with soybean mosaic virus. Mainly the feeding of certain insects, particularly the bean leaf beetle, transmits the virus. The virus is not seed transmitted. In the field, diseased plants show a mild yellow mottling on young actively growing leaves. As these leaves approach maturity the mottling becomes masked.

Fig. 5.7. Bean pod mottle.

Peanut mottle (Symptoms)

1. A mosaic of dark green and yellow areas is produced in older leaves.
2. Ring patterns of yellow patches develop on the third and fourth leaves following infection. Infected leaves pucker and curl downward at the margins.

Disease cycle

Peanut mottle is caused by the peanut mottle virus (PMV). It was first observed in 1971 and is the most common virus disease of soybeans in Georgia. The virus is seedborne in peanuts, but not in soybeans. Aphids can carry the virus from infected peanut fields or from other legumes in nearby fields. Volunteer peanut plants remaining in soybean fields are a source of inoculum for soybean infection.

Viral Diseases of Wheat

Yellow dwarf disease (Symptoms)

1. Leaf discoloration is usually the most notable early season symptom.
2. Leaves may be various shades of red to purple or pinkish-yellow to brown.
3. As an infected plant continues to grow, older leaves typically begin to die back from the tip and may feel somewhat leathery, while the new leaves begin to discolor.

4. Spring infections occur as well, but these commonly discolor only the flag leaf and do not cause significant yield reductions.

Fig. 5.8. Barley yellow dwarf virus (BYDV).

The Virus

The virus particles (virions) of Barley Yellow Dwarf Virus Disease (BYDV) are concentrated in phloem cells. They are polyhedral in shape, 21 to 26 nm in diameter. Isolates of BYDV differ serologically and in virulence, host range and vector specificity. Most strains are compatible in wheat and aphids, but some are mutually exclusive, having different compatibilities with certain aphid species. Seed, soil, sap or other insects do not transmit BYDV.

Transmission

BYDV is transmitted by more than 20 aphid species. Corn leaf aphids, English grain aphids, greenbugs and oat bird cherry aphids are among the most common vectors. Aphids acquire the virus by feeding on diseased plants for periods as short as 30 minutes but usually 12 to 30 hours. After a four-day latent period, the aphids can transmit the virus to healthy plants when feeding. After acquisition, aphids are able to transmit the virus as long as they live known as "persistent" or "circulative" virus. Symptom expression after feeding may be in one to three weeks, but the symptoms are very inconspicuous in fall infections.

There are four prominent strains of BYDV that are correlated with the particular vector:

RMV - transmitted regularly by Rhopalo-siphum maidis (corn leaf aphid)

RPV - transmitted regularly by R. padi (oat bird cherry aphid).

MAV - transmitted regularly by Macrosiphum avenae (English grain aphid)

PAV - transmitted regularly by R. padi and M. avenae.

Disease cycle

BYDV persists in cereal crops, in annual and perennial grasses, both tame and wild, and in the aphid vectors. Spread is dependent upon vector movement. In the fall, aphids move from grasses or volunteer cereals to newly planted wheat (or other winter cereals). In the spring, aphids that overwinter as adults in grasses or winter cereals become active vectors. Others that develop from eggs acquire BYDV in the spring by feeding on infected grasses and cereals during migration. Most aphids have a "winged model" (alate) form that provides more rapid movement. In feeding, they settle down to a non-wing form, with somewhat slower movement from plant to plant.

Aphid flights can be localized, or they can be disseminated for hundreds of miles when assisted by wind. The movement is associated with the development of the disease from south to north in North America. Missouri is a strategic state in the dissemination of infectious aphid vectors.

The fall infections are the most serious in wheat. Inoculated plants become systemically infected and develop symptoms within two weeks at 20°C, within four weeks at 25°C, but not at all above 30°C. In other words, cool (not cold) temperatures accentuate symptom expression.

Control measures

1. Although there are no high levels of resistance in wheat, there are some tolerant wheat varieties. This is also true of red leaf tolerance in oats.
2. The use of insecticides in the fall for control of aphids in the field, if properly timed, can reduce the incidence of BYD and increase yields, especially in respect to fall infections. The economics and efficacy of insecticides for vector control must be considered. It must be emphasized that aphids can inoculate the virus within a few hours, so the timing of the insecticide, especially with the short-duration insecticides, would have to be pinpointed better than most farmers could accomplish.
3. Fall infections can be offset from periods of high aphid activity by planting later in the fall when cooler temperatures slow aphid movements.

Wheat yellow mosaic

Wheat yellow mosaic (WYM) is also known as wheat spindle streak mosaic (WSSM). There are some similarities to wheat soil-borne mosaic in respect to symptoms and dissemination. However, the wheat spindle streak mosaic virus (WSSMV) tends to be uniformly distributed in fields rather than in pockets. Symptoms of WYM are most evident on lower leaves in early spring, and like WSMB when temperatures rise, the symptoms become masked.

Symptoms

1. Symptoms appear in early spring as yellow-green mottling, dashes and streaks on leaves.
2. Discontinuous streaks run parallel with veins and taper to form chlorotic spindles.
3. When temperatures are cool, the streaks progress to the flag leaves. Discolored areas tend to coalesce, followed by necrosis.
4. Reddish streaking at the leaf tips often precedes necrosis.
5. Some stunting and poor tillering occur in infected plants.
6. Head numbers and seed production are reduced, but kernel weight appears to be maintained.
7. The virus can cause production losses if it becomes extensive.
8. The optimum temperature for symptom development is 60°F. Above 68°F disease progress stops.
9. Symptoms typically appear in early spring right after green-up. By the time of jointing, mosaic symptoms have usually faded, but stunting may persist until maturity.

The Virus

Particles (virions) of WSSMV are rods that are normally 15 to 18 nm wide and 200 to 2,000 nm in length. The threadlike virions are rather unique and they are often clustered together, characteristics that are helpful diagnostic features for electron microscopy. The WSSMV particles usually are found in epidermal and parenchyma tissues of infected leaves. They are sparse and are difficult to isolate in leaf-dip preparations.

Fig. 5.9. Chlorotic leaf streaks parallel to the veins caused by virus infection.

The Vector

Interestingly, the wheat spindle streak mosaic virus has the same fungus vector as the wheat soil-borne mosaic virus. Polymyxa graminis harbors the virus, and the virus enters the wheat plants through root entry by the fungus.

Disease cycle

The virus survives for many years in agricultural soils even in the absence of wheat. Apparently, WSSMV can exist for a long time in association with the Polymyxa graminis fungus. The virus enters wheat plant roots via the fungus in the fall. Fall infections are most serious and account for the symptoms seen in early spring. Spring inoculations occur but are not as important. The disease is most significant under cool spring temperatures and is of little consequence under warm conditions. This relationship to cool temperatures probably has much to do with prevalence in any given season and also explains the association between occurrences and geographic locations. For example, there are more incidents of the spindle streak mosaic disease in southeast Missouri than in northern Missouri, where soil-borne mosaic seems to be more prevalent.

Control measures

1. Crop rotations also appear to limit incidence, in spite of the fact that the virus can survive for many years.
2. There are no pesticides, which provide economic control of soilborne mosaic. Liberal use of nitrogenous fertilizers such as urea and poultry manure decrease soil infectivity.
3. Any field with a history of WSBMV should be planted to varieties with resistance. Many varieties are resistant to WSBMV.
4. Late planting is sometimes effective in avoiding infection periods in the fall. Wheat planted after the "fly- free" date is less likely to be attacked by soilborne mosaic as well as other viral diseases.

Wheat streak mosaic (Symptoms)

1. WSM symptoms typically appear in April or May as yellow stunted areas along field margins. These areas are often associated with volunteer wheat from the former crop.
2. WSM symptoms become more serious with time, in contrast to the soil-borne viral diseases and the disease is much more destructive than barley yellow dwarf.

3. The yellowing and stunting become more severe and a gradual spread across the field may be seen.
4. The earliest emerging wheat, often volunteer wheat, usually is most affected.
5. Individual plants turn yellow, show definite stunting and may have wilting symptoms.
6. Tillers may be partially prostrate. Root development is often reduced.
7. Mosaic symptoms begin in younger leaves as light green to yellow dashes that enlarge to give a streaked appearance. Finally, the whole leaf blade turns yellow.
8. The younger the plants are when infected with the virus, the more severe the symptoms. Head formation can be nullified.
9. Late infections cause light and dark green streaks on the flag leaves, but not much stunting.
10. WSM goes to spring wheat, barley, corn rye, oats and a number of annual and perennial grasses (wild and tame).

The Virus

WSM is caused by the wheat streak mosaic virus (WSMV), which is carried from diseased to healthy plants by the microscopic wheat curl mite, Aceria tulipae. Particles (virions) of WSMV occur in most leaf cells as flexuous rods, 15 x 700 nm. The virus is relatively unstable and difficult to purify. It is easily transmitted through sap as well as the mites.

The Vector

The wheat curl mite, Aceria tulipae Keif, is the principal means of dissemination of WSMV in nature. There seems to be no evidence of other mites or insects that serve as vectors. Curl mites are extremely small, less than 1/100 inch in length, requiring a dissecting microscope or a 10 to 20 power hand lens for observation. The life cycle, from egg through two larval stages to the adult and egg, can be completed in seven to 10 days. Therefore, populations can build up dramatically. Under ideal conditions, one adult could theoretically produce several million offspring in 60 days. However, predators or other environmental conditions usually limit population buildups. Curl mite larvae acquire WSMV within a few minutes of feeding on infected plants. They are able to transmit the virus to healthy plants for at least a week following acquisition.

WSMV is carried in the mid and hind gut of all larvae and adult mite stages. The virus does not pass through eggs. The white and cylindrical mites possess four tiny legs next to the head that provide limited movement. They also have a tail end (anal) sucker for attachment. Their dispersal plant-to-plant and/or field-to-field is via wind. The mites spin webs that act as "sails." After landing on a suitable host, the mites crawl up to the youngest unrolled leaves in the whorl.

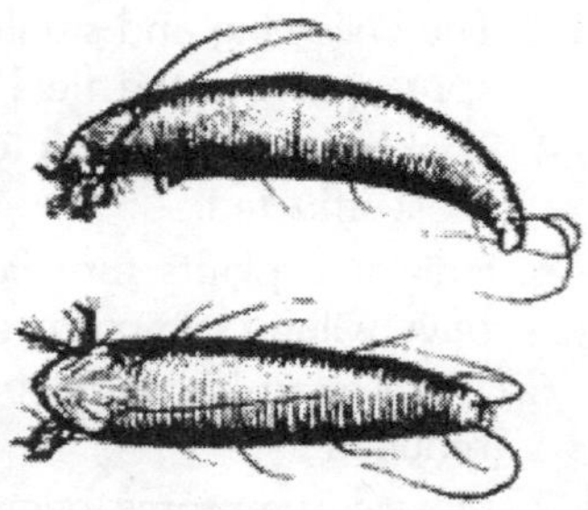

Fig. 5.10. The wheat curl mite (Aceria tulipae)

They attach themselves to the upper surface by the anal sucker and begin to feed and reproduce. Feeding prevents the leaves from unrolling normally, causing rolled leaf edges. Consequently they are known as "wheat curl mites." Heavy populations may cause enough leaf rolling so that expanding leaves and heads may be trapped. This may help to diagnose wheat curl mite infestations. As the wheat plants mature, the mites migrate upward, including the ripening heads. If grain is shattered before harvest, especially as a result of hail, the mites attached to such kernels can sometimes survive long enough to move to the sprouting seedling. In such cases, new volunteer seedlings are infected early and can be an early source for field infections. Without early emerging volunteer wheat, the wheat curl mites and the virus will die quickly following wheat maturity. Hot, dry July weather reduces mites and virus drastically. When deprived of food and water, they survive only for hours or at most two days. However, there are always mites that survive on grasses. Wet summer weather conditions permit volunteer wheat to grow, and the earlier it emerges and the curl mites become established, the higher the population. These mites can then be blown into fall seeded wheat. Warm weather in October and November extends the period of mite activity, with consequent virus inoculations. There is also some evidence that warm weather in February and March may result in further activity of overwintering mites.

Host plants

Wheat is the preferred host for curls mites and the wheat streak mosaic virus. Other cereal grains (barley, oats, triticale, rye) are slightly susceptible. Sorghum species are immune to WSMV and are poor hosts to the mites. A few corn hybrids are susceptible to the virus, and mites migrate to corn fields from mature wheat fields. Infections of corn usually do not cause serious losses, although virus symptoms can be identified. The movement of viruliferous mites from mature corn back to wheat usually is not a serious problem. The corn matures early enough that mites die before fall seeded wheat emerges. Certain summer annual grasses such as barnyardgrass, crabgrass, foxtails, sandbur and witchgrass are susceptible to the virus and can serve as potential oversummering hosts. Buffalograss, smoothbrome and western wheatgrass are perennial grasses that are immune to the virus but can support mite populations. All soft red winter wheat and hard red winter wheat varieties are more or less susceptible to WSMV. Therefore, various cultural practices must be used to minimize infection.

Disease cycle

The virus (WSMV) and the wheat curl mites persist on wheat, corn, millet and certain susceptible grasses, such as buffalograss, crabgrass and foxtail. From spring through fall, winds distribute the mites to new cereal and grass hosts. Since the virus and the mites must persist on living susceptible hosts, they are subject to extinction when wheat is ripe and gone to harvest. In mature or recently harvested fields, the mites will survive on green shoots or on volunteer plants developed from shattered grain. Volunteer wheat, therefore, becomes very important to sustaining mite populations. From these plants, which can become infected by mites feeding and reproducing, the viruliferous mites can be wind-borne into the wheat field. If weather conditions are warm, the increase of mites in the field can be great and the virus can be extensively inoculated. Fall infected wheat is the most seriously affected by the virus.

Control measutres

Control measures for wheat streak mosaic are aimed at destroying the populations of mites that transmit the WSMV and destroying the plants, which are the virus source. Taking the following measures best does this:

1. Destroy all volunteer cereals, old cereal stubble, and weed grasses in adjoining fields two weeks before planting, and three to four weeks before sowing in the field to be seeded. Doing

Table 5.1. Grass plants tested for suitability as hosts for wheat curl mite survival and wheat streak mosaic susceptibility

Common and Scientific Names	*Increase of mites*	*Mosaic Susceptible*
A. Crop Plants		
Oats (*Avena sativa*)	none	yes
Barley (*Hordeum vulgare*)	poor	yes
Rye (*Secale cereale*)	poor	yes
Sorghum (*S. vulgare*)	fair-good	no
Sudangrass (*S. vulgare var. sudanense*)	poor	no
Corn (*Zea mays*)	poor-good	yes
Foxtail millet (*Setaria italica*)	poor	yes
Broom-corn millet or proso (*Panicum miliaceum*)	none	yes
Wheat (*Triticum aestivum*)	good	yes
B. Annual Grasses		
Jointed goatgrass (*Aegilops cylindrica*)	fair-good	yes
Wild Oats (*Avena fatua*)	none	yes
Japanese chess (*Bromus japonicus*)	none	yes
Cheat (*B.secalinus*)	good	yes
Downy chess (*B.tectorum*)	none	yes
Field sandbur (*Cenchrus pauciflorus*)	good	yes
Smooth crabgrass (*Digitaria ischaenum*)	fair-good	yes
Hairy crabgrass (*D. sanguinalis*)	none	yes
Barnyard grass (*Echinchloa crus-galli*)	poor	yes
Goosegrass (*Eleusine indica*)	none	no
Stinkgrass (*Eragrostis cilianensis*)	poor	yes
Teosinte (*Euchlaena mexicana*)	poor	no
Foxtail Barley (*Hordeum jubatum*)	poor	no data
Witchgrass (*Panicum capillare*)	none	yes
Yellow foxtail or bristlegrass (*Setaria lutescens*)	none	no
Bristly or bur foxtail (*S. verticillata*)	poor	yes
Green foxtail (*S. viridis*)	poor	yes
C. Perennial Grasses		
Tall wheatgrass (*Agropyron sp.*)	none	no
Western wheatgrass (*A. smithii*)	poor-fair	no
Slender wheatgrass (*A. trachycaulum*)	none	no data

Crested wheatgrass (*A.desertorum*)	none	no data
Meadow foxtail (*Alopecurus pratensis*)	none	no
Tall oatgrass (*Arrhenatherum elatius*)	poor	no
Blue grama (*Bouteloua gracilis*)	none	no
Side-oats grama (*B. curtipendula*)	none	no
Grama (*B. sp.*)	good	yes
Smooth brome (*Bromus inermis*)	very poor	no
Buffalograss (*Buchloe dactyloides*)	none	no data
Orchardgrass (*Dactylis glomerate*)	none	no
Canada wildrye (*Elymus canadensis*)	fair	yes
Indian ricegrass (*Orzyopsis hymenoides*)	poor-fair	yes
Switchgrass (*Panicum virgatum*)	none	no
Reed canarygrass (*Phalaris arundinacea*)	none	no
Canada bluegrass (*Poa compressa*)	poor	yes
Wheeler bluegrass (*P. nervosa*)	poor-fair	no
Bulbous bluegrass (*P. bulbosa*)	poor	yes
Bulbous bluegrass (*P. stenantha*)	poor	yes
Johnsongrass (*Sorghum halepense*)	good	no
Indian grass (*Sorghastrum nutans*)	none	no
Sand dropseed (*Sporobolus cryptandrus*)	none	no data
Green needlegrass (*Stipa viridula*)	none	no data
Needle-and-thread (*S. comata*)	poor-fair	no data
Pairie sandreed (*Calamovilfa longifolia*)	none	no data

this eliminates the mite vector as well as the mosaic-infected plants. The best control results when all wheat farmers in a community cooperate in destroying volunteer wheat and old stubble well ahead of planting time.

2. Sow winter wheat as late as practical after the Hessian fly-free date or the latest recommended date to escape migrations of the mite from corn, volunteer wheat or barley, or weed grasses. If winter wheat is not up until October or later, it usually escapes severe infestation, unless fall temperatures are above normal. An infection of winter cereals by wheat streak mosaic in the spring does relatively little damage.
3. Disease reactions may vary from one locality to another and from year to year, depending on the physiologic races of the pathogens present.
4. Chemical control of the wheat curl mite has not been successful. The tightly rolled and trapped leaves provide a natural

protection for the mite, preventing contact with miticides. It is also difficult to know exactly when to apply chemicals for control.

Wheat Soil-borne mosaic (Symptoms)

Plants infected with Soil-borne Wheat Mosaic Virus (SBWMV) can show main two types of symptoms.

1. The first is leaf mottling, which appears as a light green and light yellow mosaic on the leaves. The mottling will be seen only very early in the season.
2. The second symptom is stunting to the point where the wheat plant looks like a rosette when growth begins in the spring. Under good growing conditions the infected plants may recover somewhat.
3. Fields may be uniformly diseased or more often may have spots in the field with virus symptoms.
4. Symptoms are most prominent in early spring. Warming spring temperatures tend to slow disease development and eventually the symptoms are almost completely masked.

The Virus

Soil-borne Wheat mosaic is caused by the soil-borne wheat mosaic virus (SBWMV) Particles (virions) of the soil-borne wheat mosaic virus (SBWMV) are hollow rigid rods 20 nm wide and of two lengths, 110 to 160 and 280 to 300 nm. The longer rods resemble the particles of tobacco mosaic virus. Particle lengths

Fig. 5.11. Early season stunting from SBWMV.

Fig. 5.12. Early season leaf mottling from SBWMV.

differ somewhat with strains of the virus, but both short and long rods appear to be necessary for infection.

The Vector

In nature, the way of transmission is via a soil-borne fungus, Polymyxa graminis Led. This fungus is a parasite in the roots of many higher plants, including wheat. The fungus enters roots through root hairs and epidermal cells of the roots under wet soil conditions by way of motile zoospores. The fungus carries the virus into the roots in which it colonizes. The virus is released into the plant tissues, proceeding in cell-to-cell "takeover."

Disease cycle

An unusual aspect of this disease is its mode of transmission to wheat plants. The virus is transmitted to the plant by a soil borne fungus P. graminis. The virus survives in the soil in close association with this fungus. Soils may remain infested with the fungus/virus for many years. When the fungus enters wheat roots, it transmits the virus. The fungus is a water mold and favors low, wet areas of the field, and this is usually where the disease is first seen. So the fungus infects wheat roots during cool, wet periods in the fall and possibly in the early spring. Fall infections permit the virus to elaborate to damaging proportions and predispose plants to other diseases and winter injury. SBWMV is not commonly a yield-reducing disease because higher spring temperatures inactivate the virus, and then symptoms do not appear on new leaves. Normally, spring infections occur too late to cause very much injury before warmer temperatures and crop maturation inhibit virus development. The virus and its fungus vector appear to be spread by cultivation, wind, water and other factors that cause movement of infested soil within a field. Yield reductions with SBWMV are uncommon except where extremely susceptible plants are present. Most wheat varieties are resistant to this pathogen, although that can vary.

Control measures

1. The most common and practical method of virus control is to use plant resistant wheat varieties. These varieties do not allow virus replication to occur, and the infection is stopped early.
2. Other control measures are directed at reducing the time the plants are in the field when vectors are active-thus the recommendation to plant after the fly-free date, when insect activity is reduced.
3. Systemic insecticide seed treatments have also shown some success.
4. Crop rotation and planting times may have minimal benefits.

Other virus or Virus-like diseases of wheat

Brome mosaic virus (BMV), which was described on bromegrass, is capable of affecting wheat, oats, corn, barley and rye. However, natural infections of wheat are of little economic importance. Some wheat cultivars are symptomless carriers of BMV, while others may develop streak-like mottle symptoms. Maize dwarf mosaic (MDM) is a serious disease of corn. The maize dwarf mosaic virus (MDMV) is a strain of the sugarcane mosaic virus and is, of course, most important on corn and sorghum. However, it can be transmitted to wheat by aphids. MDMV in wheat induces mild leaf mottling. It will be difficult to identify. Tobacco mosaic (TM), caused by the tobacco mosaic virus (TMV), is a prevalent and stable virus in nature with a wide host range, especially among dicotyledonous species. It can, however, infect certain grasses, among them wheat. In the case of a double infection when another wheat virus is involved, some additional symptom expression could be observed. Certain virus-like diseases caused by mycoplasma-like organisms (MLO's). Symptoms may be similar to other known viral diseases, or they may be inconspicuous. The ultra-microscopic organisms are more like bacteria than viruses. They do not have rigid cell walls, but have a fine outer membrane that allows greater

Table 5.2. Identification of Wheat Viruses.

	Barley Yellow Dwarf	*Wheat Streak Mosaic*	*Soil-Borne Mosaic*	*Spindle Streak Mosaic*	*Maize Dwarf Mosaic*
When symptoms are observed	6-8 weeks after spring growth begins-yellowing at maturation, yellow to reddish flag leaves	4-6 weeks after spring growth begins. In fall on volunteer wheat.	Early spring. 1-2 weeks after spring growth begins. Rarely in fall.	Early spring. 1-2 weeks after spring growth begins.	Rare in wheat
Pattern in field	Random circular areas	Often along edgesof fields or near volunteer wheat. Diminishing with distance.	Somewhat circular areas, especially in wet areas.	More widespread in field than soil-borne mosaic.	Random plants
Leaf symptoms	Leaf tips bright yellow on upper leaves, or reddish flag leaves - most distinct.	Bright yellow leaves with streaking patterns toward tips. Most prominent on upper leaves.	Pale yellow leaves with mosaic patterns	Yellow-green mottling dashes and streaks parallel with veins	Mosaic, mottling

		Curling of upper leaves.			
Stunting	Some-but hard to identify; fewer tillers	Severe stunting from fall infections	Some stunting but also some recovery after warm weather	Mild stunting and fewer tillers	Mild stunting
Other symptoms	Poor roots; susceptible to winter injury	Poor roots; wilting; prostrate tillers	Poor roots; winter injury	Poor roots; winter injury	Incons-picuous
Vectors	Several aphid species	Wheat curl mite *Aceria tulipae*	Fungus *Polymyxa graminis*	Fungus *Polymyxa graminis*	Several aphid species
Conditions favoring infection	Long warm fall	Long warm fall. Early volunteer wheat from hail or other factors at harvest.	Wet soil in fall. Cool temperat-ures in spring.	Wet soil in fall. Cool temperat-ures in spring.	Proximity to infected Johnson-grass
Control	Resistant or tolerant varieties	Destroy volunteer wheat. Delay planting to "fly free date."	Soft red winter wheat varieties usually more tolerant than hard red winter wheat varieties	Resistant cultivars. Late planting.	Not needed

flexibility than bacteria. Since Aster yellows has a wide host range and is very prevalent in nature, wheat can ostensibly have aster yellows infections. It is leaf hopper transmitted. Essentially, it is of minimal consequence to wheat production.

Viral Diseases of Pepper

Symptoms

1. They may also develop inward rolling and distorted tips. On fruits, solid circular spots may develop.
2. In many instances, they become rough, deformed and severe reduction in size.
3. When plants are affected when still young, they are stunted and express severe symptoms.

Disease cycle

The pepper mild mottle virus causes pepper mild mottle disease. The host range of the viruses consists of more than 750 plant species in various families including many vegetables such as tomato, cucumbers and legumes. The extensive host range serves as a source of infections and contributes to the spread of the viruses to the crop. The viruses are introduced to cultivated peppers mainly by aphid species after they have fed and acquired the viruses from wild reservoir hosts. More than sixty aphid species, including the common Aphis gossypi and Myzus persicae, are capable of transmitting viruses in a non-persistent manner.

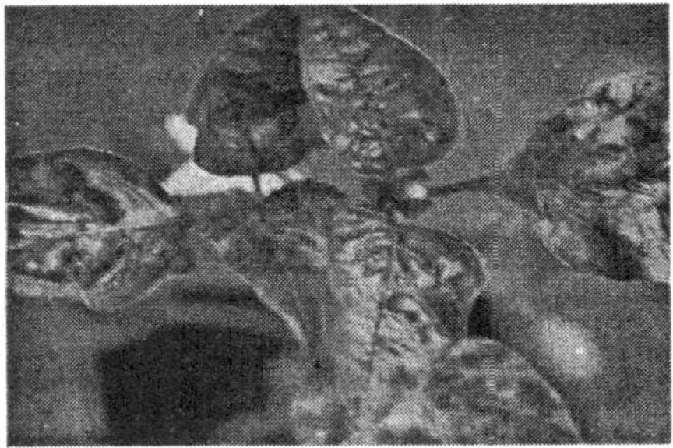

Fig. 5.13. Distorted leaves are some of the signs of viral diseases that affect pepper.

Control measures

1. The pepper should not be planted near source plants such as Solanum nigrum and hosts such as tobacco, tomato and cucumber.

2. Besides using the recommended methods such as planting disease-free seedlings and uprooting the diseased plants, the fields and nurseries should be protected from the aphids.
3. Using sticky yellow polythene sheets, which are erected vertically on the windward side of the fields and nurseries, can help ward off these vectors. The aphids are attracted to the yellow colour and are caught on the sticky polythene.
4. The most effective control method is the use of virus-resistant varieties.

Tobacco Mosaic Virus Disease of Tomato and other Plants

The plant disease caused by tobacco mosaic virus is found worldwide. The virus is known to infect more than 150 types of herbaceous, dicotyledonous plants including many vegetables, flowers, and weeds. Infection by tobacco mosaic virus causes serious losses on several crops including tomatoes, peppers, and many ornamentals. Tobacco mosaic virus is one of the most common causes of virus diseases of plants in the world. Many viruses produce mosaic-like symptoms on plants.

Symptoms

1. Mosaic-like symptoms are characterized by intermingled patches of normal and light green or yellowish colors on the leaves of infected plants. Symptoms include various degrees of chlorosis, curling, mottling mosaic, dwarfing, distortion, and blistering of the leaves.
2. Tobacco mosaic damages the leaves, flowers, and fruit and causes stunting of the plant. Many times the entire plant is dwarfed and flowers are discolored.
3. The virus almost never kills plants but lowers the quality and quantity of the crop, particularly when the plants are infected while young.
4. Virus-infected plants often are confused with plants affected by herbicide or air pollution damage, mineral deficiencies, and other plant diseases.
5. Symptoms can be influenced by temperature, light conditions, nutritional factors, and water stress.

Common Plant Hosts

Common plant hosts for the mosaic virus are tomato, pepper, petunia, snapdragon, delphinium, and marigold. Tobacco mosaic virus also has been reported to a lesser extent in muskmelon,

Fig. 5.14. Tomato leaves infected with TMV.

cucumber, squash, spinach, celosia, impatiens, ground cherry, phlox, zinnia, certain types of ivy, plantain, nightshade, and jimson weed. Although tobacco mosaic virus may infect many other types of plants, it generally is restricted to plants that are grown in seedbeds and transplanted or plants that are handled frequently.

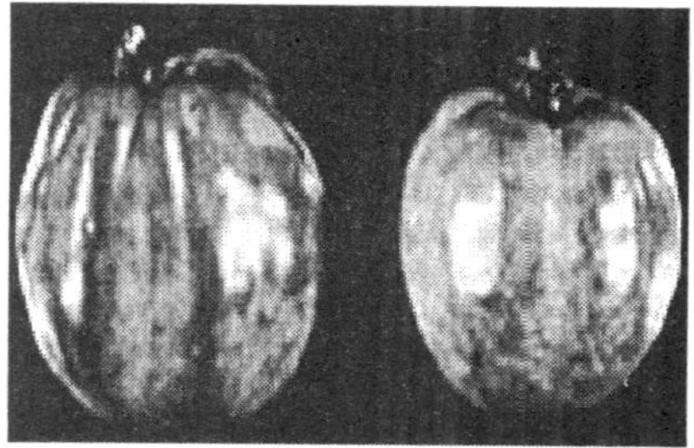

Fig. 5.15. Tomato fruit infected with TMV.

Symptoms of tomato plants

1. In tomatoes, the foliage shows mosaic (mottled) areas with alternating yellowish and dark green areas.
2. Leaves are sometimes fern-like in appearance and sharply pointed.
3. Infections of young plants reduce fruit set and occasionally cause blemishes and distortions of the fruit.
4. The dark green areas of the mottle often appear thicker and somewhat elevated giving the leaves a blister-like appearance.

The Virus and Disease cycle

Viruses differ from fungi and bacteria in that they do not produce spores or other structures capable of penetrating plant parts. Since viruses have no active methods of entering plant cells,

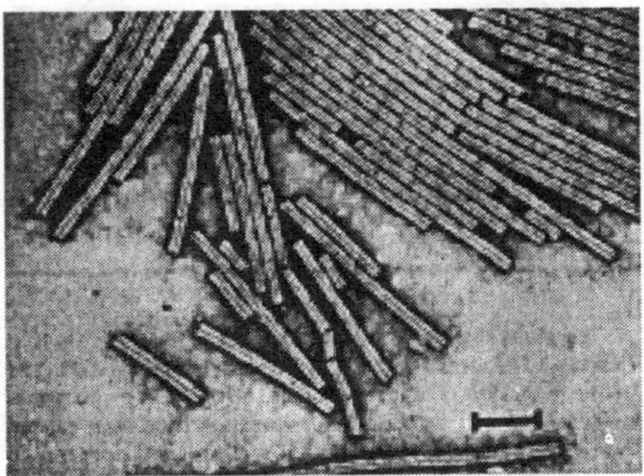

Fig. 5.16. Photograph taken with an electron microscope of the rather stiff, rod-shaped virus particles of tobacco mosaic virus from infected tomato. The bar represents 200 nanometers or 0.000008 inches.

they must rely upon mechanically caused wounds, vegetative propagation of plants, grafting, seed, pollen, and being carried on the mouthparts of chewing insects. Tobacco mosaic virus is most commonly introduced into plants through small wounds caused by handling and by insects chewing on plant parts. The most common sources of virus inoculum for tobacco mosaic virus are the debris of infected plants that remains in the soil and certain infected tobacco products that contaminate workers hands.

Cigars, cigarettes, and pipe tobaccos can be infected with tobacco mosaic virus. Handling these smoking materials contaminates the hands, and subsequent handling of plants results in a transmission of the virus. Therefore, do not smoke while handling or transplanting plants. Once the virus enters the host, it begins to multiply by inducing host cells to form more viruses. Viruses do not cause disease by consuming or killing cells but rather by taking over the metabolic cell processes, resulting in abnormal cell functioning. Abnormal metabolic functions of infected cells are expressed as mosaic and other symptoms as previously described. Infected plants serve as reservoirs for the virus and the virus can be transmitted easily (either mechanically or by insects) to healthy plants.

Control measures

1. Unlike fungicidal chemicals used to control fungal diseases, to date there are no efficient chemical treatments that protect plant parts from virus infection. Additionally, there are no known chemical treatments used under field conditions that eliminate viral infections from plant tissues once they do occur.

2. Practically speaking, plants infected by viruses remain so. Thus, control of tobacco mosaic virus is primarily focused on reducing and eliminating sources of the virus and limiting the spread by insects. Tobacco mosaic virus is the most persistent plant virus known. It has been known to survive up to 50 years in dried plant parts. Therefore, sanitation is the single most important practice in controlling tobacco mosaic virus.
3. The most common method of transferring the virus from plant to plant is on contaminated hands and tools. Workers who transplant seedlings should refrain from smoking during transplanting and wash their hands frequently and thoroughly with soap and water.
4. Tools used in transplanting can be placed in boiling water for 5 minutes and then washed with a strong soap or detergent solution. Dipping tools in household bleach is not effective for virus decontamination.
5. Any seedlings that appear to have mosaic symptoms or are stunted and distorted should be removed and destroyed. After removing diseased plants, never handle healthy plants without washing hands and decontaminating tools used to remove diseased plants.
6. Persons purchasing small tomato plants for transplanting should beware of any plants showing mottling, dwarfing, or stunting. Avoid the purchase of any affected plant.

Control method for Commercial Producers of tomato plants

Commercial greenhouse producers of tomatoes should follow control practices for seedling production as stated above. It is essential for commercial growers to constantly inspect and rogue diseased production plants while the plants are in the seedling stage. An experienced individual, who is familiar with the tobacco mosaic virus symptoms, should do the initial inspection. Roguing of young production plants is recommended and should take place before workers are allowed to prune or tie up production plants. When removing diseased plants, also remove one plant on either side of the diseased one. The reason for this is that it is almost impossible to remove a diseased plant and not contaminate the healthy adjacent plants. Never attempt to transplant a healthy tomato into the soil from which a diseased plant was removed.

Roots from diseased plants will remain in the soil and provide the virus inoculum for the new transplant. As a matter of routine, soils from which production plants have been removed, following harvest should be steam sterilized before the introduction of new seedlings. Steam or air-steam mixtures can accomplish steam sterilization. In the preparation of soil for steam sterilization, sift it to remove clumps and large pieces of organic matter. The total soil mixture will have to be heated to a temperature of 200° F for 40 minutes. Since high temperatures are required, steam sterilization must be done in an enclosed system. High temperature thermometers to make sure the desired temperature has been reached should monitor temperatures within the steam sterilization system. Steam sterilization of soil also will eliminate fungi, insects, nematodes, and weeds from the soil. Steam sterilization also is recommended for gravel mixtures used in hydroponic operations following the same procedure described above. Grow individual production plants in separate containers so that the soil or growing media can be removed when roguing infected production plants. Remember that the soil harbors old root tissues that may serve as inoculum when new roots are introduced. Growing production plants in separate containers is also useful for the control of root diseases caused by fungi and bacteria.

Plum Pox Potyvirus (PPV) Disease of Stone Fruits

Symptoms

1. In Prunus, plum pox virus (PPV) symptoms appear on leaves, fruits, flowers, and seeds.
2. The severity of the symptoms varies according to the Prunus species and cultivar, PPV strain, season and location.
3. Leaves and fruit show chlorotic (yellowing) and necrotic (browning) ring patterns, and chlorotic bands or blotches.
5. Leaves and fruit also can be absent of symptoms, or have symptoms that are ameliorated during the growing season.
6. The fruit of apricot and plum can be misshapen and deformed or rings may be present on their stones.
7. Some peach cultivars may show color-breaking symptoms on the flower petals.
8. Virus infection can cause considerable losses. About 100 million stone fruit trees in Europe are currently infected, and susceptible cultivars can result in 80-100% yield losses (Kegler,

1998). In eastern and central Europe, sensitive plum varieties can exhibit premature fruit drop and bark splitting.

9. Some sweet cherry fruits develop chlorotic and necrotic rings, notched marks, and premature fruit drop.

Table 5.3. Plum Pox Status and Associated Geographic Distribution

Disease Status	*Country*
Restricted Distribution	Albania, Austria, Cyprus, Czech Republic, France, Italy, Luxembourg, Moldova, Norway, Portugal, Southern Russia, Slovenia, Spain, Syria, Turkey, Ukraine, United Kingdom, United States
Widespread	Bulgaria, Croatia, Germany, Greece, Hungary, Poland, Romania, Slovakia, Former Yugoslavia
Introduced, Established	Azores, Bosnia-Herzegovina, Egypt, Former Russia, India, Lithuania
Introduced, Presumably Eradicated	Belgium, Netherlands, Switzerland
Present Status Unknown	Chile, Denmark

The Virus

Plum pox virus belongs to the genus Potyvirus in the family Potyviridae. Members of the genus Potyvirus have virions which are flexuous filaments with no envelopes, are aphid-transmitted in a non-persistent, stylet-borne manner, mechanically transmitted, and may or may not be seed-transmitted.

PPV has a single molecule of positive sense, ssRNA, that is 9.7 kb; virions are approximately 764 X 20 nm. The genome is expressed as a 350 kDa polyprotein precursor that is proteolytically processed by viral and host proteases into seven smaller functional proteins including a 3' coat protein and a helper component. It is the only recognized potyvirus infecting Prunus. The introduction of PPV to a new country or region is usually through propagative materials and the subsequent distribution of contaminated materials. The secondary spread can be rapid and results from aphid transmission.

Disease cycle

Plum pox virus has been transmitted by at least 20 aphid species, although only 4-6 are considered important vectors. The efficiency of transmission is dependent on the virus strain, host cultivars, age of the host cultivars, aphid species, and time of year. The most important aphid vectors reported from several countries are Brachycaudus cardui, B. helichrysi, Myzus persicae, and Phorodon humuli. Reports vary from country to country, however, natural virus spread is low in July and August but high in spring and autumn. Spring flights of B. helichrysi, M. persicae, and P. humuli are most important for spread within and between orchards. Analysis of spacial distribution of PPV by Gottwald et al. (1995) suggest a lack of movement by aphid vectors to immediately adjacent trees and a preference for movement several tree spaces away. Aphids can acquire the virus in probes as short as 30 seconds, and can transmit for up to 1 hour. Aphids that have been starved before feeding can transmit for up to 3 hours after acquisition. There is no correlation between the ability to transmit PPV and the ability to colonize Prunus. PPV can be spread in orchards by transient aphids as efficiently as aphids colonizing Prunus (Labonne et al., 1995). Aphids were found to transmit PPV within 100-120 m of the source plants, but they have been shown to carry the virus on their stylets for several kilometers if starved during flight.

The plant host

Plum pox virus has a broad experimental host range although it has a rather restricted natural host range within the genus Prunus. It infects peaches, plums, apricots, nectarines, almonds, and sweet and tart cherry. Virus isolates vary in their reaction to different hosts, and not all strains or isolates infect the same host range. Prunus species that have been proven to be hosts in nature, or by inoculation trials followed by back transmissions include:

Prunus armeniaca - Apricot

P. persica - Peach

P. persica var. nectarina - Nectarine

P. domestica - Garden plum (prune)

P. salicina - Japanese plum

P. insititia - Damson plum

P. cerasifera - Myrobalan plum

P. glandulosa - Dwarf flowering almond, Cherry almond

P. avium - Sweet cherry

P. cerasus - Sour (tart) cherry

P. amygdalus - Almond

Wild Prunus may serve as an important secondary host of PPV and can have an impact on plum pox epidemiology and control (Polak, 1997). In addition to the above natural hosts, several wild Prunus species are susceptible:

P. spinosa - Blackthorn

P. americana - American plum

P. bessey - Western sand cherry

P. mahaleb - Mahaleb or St. Lucie cherry

P. mume - Japanese apricot

P. pumila - Sand cherry

P. hortulana - Hortulan plum

P. davidana - David peach, Chinese wild peach

P. tomentosa - Nanking cherry

P. nigra - Canada plum

P. maritime - Beach plum

P. laurocerasus - English cherry-laurel

Many non-Prunus species, in at least nine plant families, have been infected artificially with one or more strains of the plum pox virus, and in some cases found naturally infected in the field. Most of these are herbaceous annuals but a few are perennial or woody and could serve as over-wintering sources of the virus. Some of the common hosts include:

Campanula rapunculoides

Chenopodium quinoa

C. species

Lamium album

L. amplexicaule

L. purpureum

Lupinus albus

Lycium barbarum

L. halimifolium

Medicago lupulina

Melilotus officinalis

Ranunculus acer

R. arvensis

R. repens

Silene vulgaris

Solanum dulcamara

Trifolium incarnatum

T. pratense

T. repens

Zinnia elegans

Z. violacea

In addition the following are the more important herbaceous indicator or propagation hosts of plum pox virus:

Chenopodium foetidum

Nicotiana benthamina

N. bigelowii

N. clevelandii

N. occidentalis

N. edwardsonii

N. megalosiphon

N. tabacum

N. physalodes

Pisum sativum cv. Colmo

The virus is unevenly distributed in trees that are newly infected or have some degree of resistance, however, once an infection is established it can reach high titers in plant tissues such as leaves, flowers, and fruit in the spring and early summer. Several polyclonal and monoclonal antibodies have been developed and are used worldwide for PPV detection.

Control measures

There is no specific control measures are available at present to control this disease.

Viral Diseases of Potato

(a) Leaf roll of potato: (Symptoms)

1. The margins or the tips of the leaves become yellow.
2. The leaves of the infected plants roll up along the margins, leaving the midrib at the bottom of the trough.

3. In plants with diseased seed tubers, this rolling of leaves starts in the lower leaflets and progresses upward throughout the entire plant.
4. In some varities individual leaflets have a tendency to be more erect, giving the diseased plant an upright appearance.
5. As a result of infection, necrosis occurs in the phloem and there is excessive accumulation of starch.
6. The number of tubers produced per plant and their sizes in a diseased crop are greatly reduced.

Disease cycle

This disease caused by leaf roll virus also known as potato virus I or Solanum virus 14. Virus is not transmitted through the sap. Transmission of the virus in nature occurs through infected seed tubers and through the agency of insects. The main insect vector is an aphid Myzus persicae. The virus persists in the body of Myzus persicae for many days even if the vector has fed on many immune hosts after feeding on the diseased plants.

Host plants

Datura stramonium
Datura tulula
Physalis angulata
Physalis floridana
Lycopersicon esculentum

Control measures

1. Potato seed certification and roguing are recommended for disease control.
2. Some results of experiments suggest that if the tubers are stored at 37.5 C in a humid atmosphere for 20-25 days, they become free of the leaf roll virus.

(b) Mild mosaic of potato: (Symptoms)

1. Although the disease occurring widely in most of the seed stocks, causes very slight or negligible symptoms of interveinal mottling or mosaic.
2. Under favourable growth conditions of the host of the symptoms are masked, thus making recognition of the disease difficult.
3. The mosaic symptoms are almost completely masked at high temperatures (above 21^{0}C).

4, The virus is carried by the plants without any visible symptoms in many cells.
5. There is only slight dwarfing of the plant or deformation of the foliage.
6. The virus produces top necrosis in some verities.

Disease cycle

Mild mosaic of potato is caused by potato latent virus, potato mottle virus, potato virus X or Solanum virus I. The virus contains a large number of strains. The particles of the virus are long., flexuous rods (515 mu). The virus is easily transmitted by sap inoculation and is one of the few viruses, which spreads in the feild by the contact between healthy and diseased plants. Virus can be transmitted by core grafting and by dodder, Cascuta campestris. No insect vector is known. Hower, the source of perennation and spread in the feild under natural conditions is mostly the diseased seed stock.

(c) Rugose Mosaic of Potato: (Symptoms)

1. The disease causes very severe damage to individual plants.
2. The foliage is not only mottled but is also severely wrinkled, puckered and markedly reduced in size.
3. The leaflet margins are rolled downwards and the entire plant is dwarfed.
4. The lower leaves generally have black necrotic veins.
5. While the symptoms of mottle may be masked by high temperature, roughness or rugosity and abnormal hairyness of leaves and dwarfing of the entire plant persist.
6. In severe cases plants may even die before producing any tuber.

Disease cycle

The disease ia caused by a combination of two viruses potato virus X (mild or latent mosaic) and potato virus Y. The latter causes blighting of the leaves in severe attacks and is also known as vein banding virus or Solanum virus 2. Under field condition of transmission of the virus is through small sized tubers. On diseased plants Myzus persiae is capable of efficiency transmitting virus Y which is one of the components of the complex.

Control measures

1. Systemic rouging of diseased plants as soon as they are noticed and their destruction by burring is recommended.

2. Indirect rouging of diseased seeds tubers is recommended at planting time by using large sized potatoes.

(d) Crinkle of Potato: (Symptoms)

1. The disease resembles rugose mosaic very closely but the yellowish patches on the foliage are bigger and hence, more prominent.
2. As the plant begins to die this colour becomes more pronounced and is accompanied by rusty spots, beginning near the tip of the leaves.
3. The foliage is brittle and is easily injured.

Viral Diseases of Papaya

(a) Leaf Curl of Papaya: (Symptoms)

1. Crinkling and curling of the leaves accompanied by vein clearing and reduction in the size of the leaves are the marked symptoms.
2. Leaves become leathery and brittle and the interveinal areas are raised on the upper surface due to hypertrophy, which produces rugosity.
3. The most prominent symptom is the downward and inward rolling of leaves and the veins get thickened and turn dark green in colour.
4. Sometimes all the leaves at the top of the plant areget affected by these symptoms
5. The petioles are twisted in a zigzag manner.
6. The severely affected plants fail to flower or bear fruits.
7. In the advanced stages of the diseases, defoliation takes place and plant growth is stunted.

Disease cycle

Tobacco virus 16 or Nicotiana virus causes this disease. Virus is not transmitted. It can be readily transmitted by grafting or juice inoculation. In nature the most important agent of its transmission is the white fly Bemisia tabaci.

Host plants

Papaya
Tomato
Tobacco
Zinnia
Althaea

Control measures

No suitable control method is known. Roguing is helpful in reducing the incidence of the disease.

(b) Mosaic of Papaya: (Symptom)

1. The disease is characterized by mosaic symptoms on the leaves with blister-like patches of green tissue distributed all over the yellowish green lamina.
2. Sometimes dark green spots and elongated streaks appearing like water-soaked areas, are formed on the petiol and the stem.
3. Fruits generally remain much smaller and get deformed. On some fruits large mosaic patches are also formed.
4. In severe infections, the plant has only a few small chlorotic or tendril-like leaves and they fail to flower.

Fig. 5.17. Papaya leaf roll.

Disease cycle

The papaya mosaic disease is caused by papaya mosaic virus.This virus is not seed borne. It can be transmitted by mechanical sap inoculation and wedge and bud grafting of aphids. Aphis malvae, Aphis gossypi and Aphis medicaginis are known as vectors. It is transmitted under natural conditions only through vectors.

Host plants

Trichosanthes anguina
Cucurbita maxima
Cucurbita pepo

Citrullus vulgaris
Cucumis sativus
Luffa acutangula
Lagenaria ciceraria

Viral Diseases of Cucurbits

Symptoms

1. Mosaics and stunting, reduced fruit yield.
2. Reduction of leaf laminae ("fernleaf").
3. Severe chlorosis, chlorotic local lesions, necrotic local lesions.

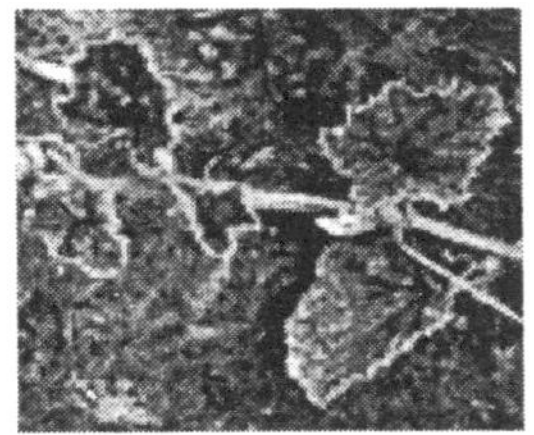

Fig. 5.18. Reduction of leaf laminae.

Fig. 5.19. Mosaic of cucurbits leaf.

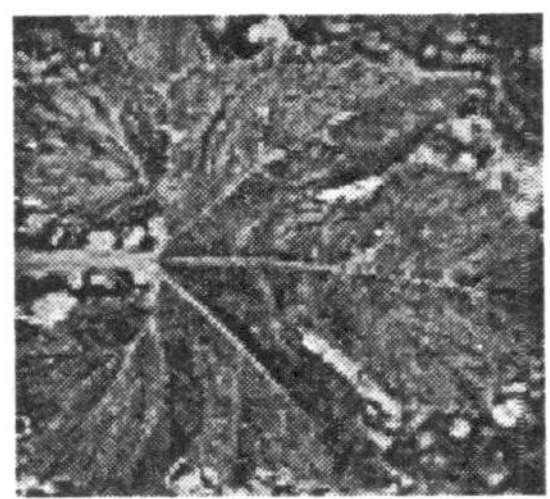

Fig. 5.20. Stunting of leaf laminae.

Host plants

Cucumis sativus
Lycopersicon esculentum
Spinacia oleracea
Chenopodium amaranticolor
Vigna unguiculata-.
Nicotiana edwardsonii
N. glutinosa
N. tabacum

Disease cycle

This disease is caused by cucumovirus. It is transmitted by a vector; an insect; more than 60 spp. including Acyrthosiphon pisum, Aphis craccivora and Myzus persicae; Aphididae. Transmitted in a non-persistent manner. Virus transmitted by mechanical inoculation; transmitted by seed (in 19 species, but in variable extents).

Synonyms of virus

Banana infectious chlorosis virus, coleus mosaic virus (Creager, 1945; Holcomb and Valverde, 1991), cowpea banding mosaic virus, cowpea ringspot virus, cucumber virus 1, lily ringspot virus (Brierley and Travis, 1958), pea top necrosis virus, peanut yellow mosaic virus, southern celery mosaic virus (Doolittle, 1916; Price, 1935; Wellman, 1934), soybean stunt virus (Hanada and Tochihara, 1982), spinach blight virus, tomato fern leaf virus, pea western ringspot virus.

Viral Diseases of Okra

(a) Okra mosaic: (Symptoms)

1. Systemic symptoms appear on the youngest leaves as light green and regular veinal chlorosis.
2. Symptoms appear on two or three subsequent leaves but leaves produced later do not show any symptom.

Disease cycle

This disease is caused by Okra mosaic virus. This is reported from West Africa only. It belongs to tymoviruses group with ssRNA and is transmitted by sap inoculation but no vector is known.

(b) Yellow vein mosaic: (Symptoms)

1. The main symptom of this disease is vein clearing and veinal chlorosis of leaves.
2. The yellow network of veins is very conspicuous and the veins and the veinlets are thickened.
3. In severe cases the chlorosis may extend to the interveinal areas and may result in complete yellowing of the leaf.
4. Fruits are dwarfed, malformed and yellowing green in colour.

Disease cycle

The virus responsible for disease is Hibiscus virus Ior Yellow vein mosaic virus. It is transmitted by whitefly (*Bemisia tabaci)* and the virus is not sap transmissible. But under artificial conditions it could be transmitted by grafting.

Host plants

Croton sparsiflora
Malvastrum tricuspidatum
Ageratum sp.

Control measures

1. Protection of the crop from whiteflies and other insect vectors by spraying with Follidol (0.3 %) or other suitable insecticides reduces the incidence of the disease.
2. Destruction of weed hosts should also be given importance.
3. At present, no definite control measures are known. Variety of Sawani is reported to be tolerant.

Viral Diseases of Sugarcane

Edgerton (1968), Martin et al. (1961) and Hughes, et al. (1964) have described the following diseases of sugarcane as viral diseases:

(a) Mosaic
(b) Fiji
(c) Sereh
(d) Chorotic streak
(e) Streak
(f) Australian dwarf
(g) Dwarf
(h) Grassy Shoot
(i) Q.13 disease
(j) Ring mosaic
(k) Striate mosaic
(l) Sembur
(m) Spike

(a) Mosaic of Sugarcane: (Symptoms)

1. The first symptom of this disease appears about six weeks after planting and continue to develop throught the mansoon after which they become obliterated on maturity of the plant.
2. The primary and critical symptom of the disease is the appearance of pale patches or blotches in the green tissues of the leaf.
3. Small areas of the leaf are of a paler green colour than the rest. These patches are not uniform in size and shape.

4. The youngest unfolded leaves show the mottling very clearly while the symptoms are not very clear on older leaves.
5. Sometimes leaves of young tillers are found to be stiff, erect and crinkled.
6. Mottling of the stem also occurs in some varities and may lead to death of cells resulting in formation of cankers.

Disease cycle

The disease is caused by Sugarcane mosaic virus which belongs to potyviruses group. Its synonyms are Grass mosaic virus, *Saccharum* virus 1 and *Marmor sacchari*. The virus is transmitted from sugarcane to sugarcane by atleast seven species of aphids such as *Dactynotus ambrosiae, Hysteroneura setariae, Rhopalosiphum maidis, Toxoptera graminum, R. maidis,* and *Shizaphis graminum* are reported as vectors of sugarcane mosaic virus. Transmission is in the non- persistent manner.

Host plants

Maize
Sugarcane
Sorghum
Saccharum spontaneum

Control measures

1. Use only selected healthy setts for seed.
2. Eliminate the weed hosts.
3. Resistant varities are recommended for controlling the disease.

Other Viruses and their Diseases

Alfalfa Mosaic Virus (AMV)

Symptoms: Plants are mildly stunted and have whitish blotches on the leaves. The infected area is white-bleached and mottled. The fruit may be distorted.

Control: Aphids carry AMV. Peppers planted near alfalfa have a higher incidence of the disease. Control aphids and avoid planting near alfalfa. Use resistant varieties if they are available.

Beet Curly Top Virus (BCTV)

Symptoms: The most striking symptom is stunted plants and yellowing of the plants. The plants are also quite stiff and erect, and the leaves have a leathery teel.

Control: Curly top is carried by leafhoppers. Help prevent losses by partially shading plants with muslin tents or by other means early in the season. Leafhoppers that carry the virus do not feed in shady locations. Spraying or dusting with insecticide can be justified only when control is needed for other insects.

Cucumber Mosaic Virus (CMV)

Symptoms: Aphids carry CMV and cause stunted plants with dull green foliage with a leathery appearance.

Control: Control aphids and avoid planting near cucurbits.

Pepper Mottle Virus (PeMV)

Symptoms: Stunted plants, distorted fruit, and yield reduction are symptoms.

Control: Aphids carry PeMV. Control aphid vectors, practice good sanitation, and plant resistant varieties if available.

Potato Y (PVY)

Symptoms: Symptoms include mosaic and dark green vein-banding, leaf crinkle, leaf distortion, and plant stunting.

Control: PVY is carried by aphids. This has been called the most common pepper virus. Plant resistant varieties and control aphids.

Tobacco Etch Virus (TEV)

Symptoms: Mosaic and dark green vein-banding, leaf distortion, and plant stunting are symptoms. Tabasco plants wilt and die.

Control: Aphids carry TEV. Plant resistant varieties and control aphids.

Samsun Latent Tobacco Mosaic Virus (SLTMV)

Symptoms: Typical symptoms include mild mosaic and leaf distortion. Pods develop rings, line patterns, necrotic spots, and distortion. Plant stunting may also occur.

Control: Hands touching an infected plant and then touching an uninfected plant, so disinfecting hands with alcohol helps spread SLTMV mechanically. Clean seed and crop rotation also help control the virus.

Tomato Spotted Wilt Virus (TSWV)

Symptoms: TSWV is carried by thrips that feed on various virus-infected perennial flowering plants commonly grown around homes. Necrotic ringspots and leaf drop are symptoms.

Control: Use clean seed and control thrips.

Black Streaked Dwarf Virus (BSDV)

Symptoms: Stunting or dwarfing occurs, particularly when plants are infected in the seedling stage. Symptoms on pearl millet are not well-documented. On maize, white, waxy swellings occur on veins. Foliage is dark-green with chlorotic streaks and splitting of leaf margins.

Pathogen and disease characteristics: This belongs to Reoviridae fijivirus group. ISometric particles are 75-80 nm in diameter. Transmission occurs persistently by plant hoppers.

Host plants

Pearl millet

corn

rice

finger millet

barnyardgrass

Isachne globosa

barley

wheat

barnyard millet

Digitaria sanguinali

Geographic distribution: South Korea, Japan (on maize)

Alternative names for the disease

Rice black-streaked dwarf

Maize streaked dwarf

Seed transmission: Probably not seed transmitted.

Guinea Grass Mosaic Virus (GGMV)

Symptoms: Young diseased plants show lines of light green eye spots or a pale green mosaic, depending on cultivar. Symptoms develop into a striped mosaic by elongation and coalescing of the eye-spots. Some plants show severe symptoms with dwarfing.

Pathogen and disease characteristics: The Potyvirus is transmitted by aphids (*Hysteroneura setariae* and *Rhopalosiphum maidis*), probably non-persistently, but can be mechanically transmitted. Symptoms appear about 8 to 10 days after inoculation. Strain/host specificity may exist.

Host plants

Pearl millet

Bromus commutatus

Bromus macrostachys
Panicum crusgalli
Panicum maximum
Sorghum aroundinaceum
Zea mays
Paspalum conjugatum
Panicum maximum
Brachiaria brizantha
Brachiaria decumbens
Brachiaria dictyoneura
Brachiaria humidicola
Brachiaria jubata
Brachiaria ruzizensis
Brachiaria deflexa
Bromus arvensis
Bromus racemosus
Bromus sterilis
Coix lacryma-jobi
Echinochloa crus-galli
Oplismenus hirtelus
Panicum bulbosum
Panicum miliaceum
Paspalum racemosum
Setaria glaucum
Setaria italica
Setaria macrochaeta
Setaria verticillata
Stenotaphrum secondatum

Geographic distribution: Ivory Coast (from pearl millet)

Seed transmission: Not known to be seed transmitted.

Indian Peanut Clump Virus (IPCV)

Symptoms: Not available in the literature.

Pathogen and disease characteristics: The furovirus is vectored by soil borne fungus Polymyxa spp.

Host plants

Pearl millet
peanut

finger millet (Eleusine coracana)

foxtail millet (Setaria italica)

Geographic distribution: Natural distribution on pearl millet is not known. Occurrence has been confirmed in India.

Seed transmission: Very low rate of seed transmission (0.9%) has been observed in plants that had been grown in an infested field in India.

Panicum Mosaic Virus

Symptoms: Symptoms on pearl millet are expressed as a mild chlorotic mottle. On switchgrass, stunting can be severe in susceptible plants. Mild green mosaic and mottling, Yellow or light green blotchy mottling, mosaic, and streaking of leaves are characteristic. The entire plant or sectors of it can become chlorotic if badly stunted.

Pathogen and disease characteristics: 109S isometric virus, 28-30 nm in diameter. Single RNA (28S) and protein species (28,000 daltons). Six serotypes have been differentiated. A serological relationship exists between PMV and members of the phleum mottle virus group.

The virus is mechanically transmitted. PMV is a warm temperature virus. Incubation periods (7-18 days) are generally shorter at warmer, and longer at cooler temperatures. Optimum symptoms develop on many hosts with temperatures of 29 to 35^{0}C. Virus will remain infective in dessicated leaf tissue for up to 9 years.

Host plants

Switchgrass (Panicum virgatum L.),

broomcorn millet (P. miliaceum L.),

ticklegrass (P. capillare L.),

panicgrass (P. scribnerianum Nash),

Hall's panicum (P. hallii Vasey),

foxtail millet [Setaria italica (L.) Beauv],

barnyard grass [Echinochloa crusgalli (L.) Beauv.],

crabgrass [Digitaria sanguinalis (L.) Scop.]

Panicum ramosum,

P. decompositum,

P. turgidum,

Setaria verticillata

Setaria lutescens
Maize (Zea mays),
Panicum dichtomiflorum
Panicum virgatum L.
St. Augustinegrass
Stenotaphrum secundatum

Geographic distribution: Kansas. St. Augustine decline stain (PMV-SADV) occurs in Texas, Louisiana, Arkansas, South Carolina, and Mexico.

Synonyms: St. Augustine decline virus (SADV) is a strain of of PMV.

Seed transmission: Generally not known to be seed transmitted, however seed transmission of an SAD strain was reported in Setaria italica.

Satellite Panicum Mosaic Virus

Symptoms: Co-inoculation of Panicum Mosaic Virus (PMV) with its satellite virus (SPMV) results in a severe chlorotic mottle on pearl millet.

Pathogen and disease characteristics: The virus is a mechanically transmitted 42S isometric particle, 17 nm in diameter. Not infectious alone. Contains two RNA species (14 and 34S) and a single protein species (15,500 daltons). Serologically unrelated to panicum mosaic virus (PMV), but dependent upon PMV for replication. Two serotypes are known.

Host range: Host range is not well defined, but probably identical to that of panicum mosaic virus.

Geographic distribution: Kansas.

Seed transmission: Not known to be seed transmitted

Table 5.4. List of some Phytopathogenic virus.

S.No.	Name	Taxon
1.	Cucumber mosaic virus	Cucumovirus
2.	Turnip yellow mosaic virus	Tymovirus
3.	Brome mosaic virus	Bromovirus
4.	Potato virus X	Potexvirus
5.	Prunus necrotic ringspot virus	Ilarvirus
6.	Raspberry ringspot virus	Nepovirus
7.	Carnation mottle virus	Carmovirus
8.	Turnip mosaic virus	Potyvirus

9.	Lettuce mosaic virus	Potyvirus
10.	Cacao swollen shoot virus	Badnavirus
11.	Cacao yellow mosaic virus	Tymovirus
12.	Tobacco rattle virus	Tobravirus
13.	Beet yellows virus	Closterovirus
14.	Tobacco necrosis virus	Necrovirus
15.	Satellite virus	TNV satellite subgroup
16.	Arabis mosaic virus	Nepovirus
17.	Tobacco ringspot virus	Nepovirus
18.	Tomato ringspot virus	Nepovirus
19.	Prune dwarf virus	Ilarvirus
20.	Echtes Ackerbohnemosaik virus	Comovirus
21.	Carnation ringspot virus	Dianthovirus
22.	Red clover vein mosaic virus	Carlavirus
23.	Cocksfoot mottle virus	Sobemovirus
24.	Cauliflower mosaic virus	Caulimovirus
25.	Pea enation mosaic virus	Enamovirus
26.	Lettuce necrotic yellows virus	Cytorhabdovirus
27.	Cymbidium mosaic virus	Potexvirus
28.	Grapevine fanleaf virus	Nepovirus
29.	Broad bean stain virus	Comovirus
30.	Apple chlorotic leafspot virus	Trichovirus
31.	Apple stem grooving virus	Capillovirus
32.	Barley yellow dwarf virus	Luteoviridae
33.	Citrus tristeza virus	Closterovirus
34.	Wound tumor virus	Phytoreovirus
35.	Potato yellow dwarf virus	Nucleorhabdovirus
36.	Potato leafroll virus	Polerovirus
37.	Potato virus Y	Potyvirus
38.	Tomato black ring virus	Nepovirus
39.	Tomato spotted wilt virus	Tospovirus
40.	Bean yellow mosaic virus	Potyvirus
41.	White clover mosaic virus	Potexvirus
42.	Tulare apple mosaic virus	Ilarvirus
43.	Squash mosaic virus	Comovirus
44.	Tobacco streak virus	Ilarvirus
45.	Narcissus mosaic virus	Potexvirus
46.	Alfalfa mosaic virus	Alfamovirus
47.	Cowpea mosaic virus	Comovirus
48.	Wheat streak mosaic virus	Tritimovirus

49.	Cowpea chlorotic mottle virus	Bromovirus
50.	Celery mosaic virus	Potyvirus
51.	Dahlia mosaic virus	Caulimovirus
52.	Belladonna mottle virus	Tymovirus
53.	Beet mosaic virus	Potyvirus
54.	Potato virus A	Potyvirus
55.	Tobacco etch virus	Potyvirus
56.	Papaya mosaic virus	Potexvirus
57.	Southern bean mosaic virus	Sobemovirus
58.	Cactus virus X	Potexvirus
59.	Cocksfoot streak virus	Potyvirus
60.	Potato virus S	Carlavirus
61.	Carnation latent virus	Carlavirus
62.	Sowthistle yellow vein virus	Nucleorhabdovirus
63.	Watermelon mosaic virus	Potyvirus
64.	Sowbane mosaic virus	Sobemovirus
65.	Robinia mosaic virus	Cucumovirus
66.	Potato spindle tuber 'virus'	Pospiviroid
67.	Rice tungro virus	Waikavirus - Rice tungro spherical virus
68.	Barley stripe mosaic virus	Hordeivirus
69.	Tomato bushy stunt virus	Tombusvirus
70.	Plum pox virus	Potyvirus
71.	Tulip breaking virus	Potyvirus
72.	Maize rough dwarf virus	Fijivirus
73.	Bean common mosaic virus	Potyvirus
74.	Red clover mottle virus	Comovirus
75.	Poplar mosaic virus	Carlavirus
76.	Narcissus yellow stripe virus	Potyvirus
77.	Soil-borne wheat mosaic virus	Furovirus
78.	Carnation vein mottle virus	Potyvirus
79.	Tomato aspermy virus	Cucumovirus
80.	Cherry leaf roll virus	Nepovirus
81.	Broad bean wilt virus	Fabavirus
82.	Cucumber necrosis virus	Tombusvirus
83.	Apple mosaic virus	Ilarvirus
84.	Papaya ringspot virus	Potyvirus
85.	Broccoli necrotic yellows virus	Cytorhabdovirus
86.	Ryegrass mosaic virus	Rymovirus
87.	Potato virus M	Carlavirus

88.	Sugarcane mosaic virus	Potyvirus
89.	Beet western yellows virus	Polerovirus
90.	Cassava common mosaic virus	Potexvirus
91.	Parsnip mosaic virus	Potyvirus
92.	Peanut stunt virus	Cucumovirus
93.	Soybean mosaic virus	Potyvirus
94.	Maize mosaic virus	Nucleorhabdovirus
95.	Henbane mosaic virus	Potyvirus
96.	Lily symptomless virus	Carlavirus
97.	Pokeweed mosaic virus	Potyvirus
98.	Potato aucuba mosaic virus	Potexvirus
99.	American wheat striate mosaic virus	Cytorhabdovirus
100.	Rice transitory yellowing	Rhabdoviridae
101.	Broad bean mottle virus	Bromovirus
102.	Rice dwarf virus	Phytoreovirus
103.	Grapevine chrome mosaic virus	Nepovirus
104.	Pepper veinal mottle virus	Potyvirus
105.	Wild cucumber mosaic virus	Tymovirus
106.	Black raspberry latent virus	Ilarvirus
107.	Cocksfoot mild mosaic virus	Sobemovirus
108.	Bean pod mottle virus	Comovirus
109.	Turnip crinkle virus	Carmovirus
110.	Chrysanthemum virus B	Carlavirus
111.	Clover yellow mosaic virus	Potexvirus
112.	Pea streak virus	Carlavirus
113.	Scrophularia mottle virus	Tymovirus
114.	Hydrangea ringspot virus	Potexvirus
115.	Eggplant mottled dwarf virus	Nucleorhabdovirus
116.	Iris mild mosaic virus	Potyvirus
117.	Hippeastrum mosaic virus	Potyvirus
118.	Agropyron mosaic virus	Rymovirus
119.	Sugarcane Fiji disease virus	Fijivirus
120.	Pea early-browning virus	Tobravirus
121.	Radish mosaic virus	Comovirus
122.	Passionfruit woodiness virus	Potyvirus
123.	Oat blue dwarf virus	Marafivirus
124.	Eggplant mosaic virus	Tymovirus
125.	Turnip rosette virus	Sobemovirus
126.	Strawberry latent ringspot virus	Nepovirus
127.	Elderberry latent virus	Carmovirus

128.	Okra mosaic virus	Tymovirus
129.	Parsnip yellow fleck virus	Sequivirus
130.	Pelargonium flower break virus	Carmovirus
131.	Clover yellow vein virus	Potyvirus
132.	Chicory yellow mottle virus	Nepovirus
133.	Maize streak virus	Mastrevirus
134.	Cowpea aphid-borne mosaic virus	Potyvirus
135.	Rice black streaked dwarf virus	Fijivirus
136.	Carnation necrotic fleck virus	Closterovirus
137.	Carrot mottle virus	Umbravirus
138.	Potato mop-top virus	Pomovirus
139.	Elm mottle virus	Ilarvirus
140.	Cowpea mild mottle virus	Carlavirus
141.	Peanut mottle virus	Potyvirus
142.	Mulberry ringspot virus	Nepovirus
143.	Barley yellow mosaic virus	Bymovirus
144.	Beet necrotic yellow vein virus	Benyvirus
145.	Oat mosaic virus	Bymovirus
146.	Pea seed-borne mosaic virus	Potyvirus
147.	Bearded iris mosaic virus	Potyvirus
148.	Saguaro cactus virus	Carmovirus
149.	Rice yellow mottle virus	Sobemovirus
150.	Peach rosette mosaic virus	Nepovirus
151.	Tobacco mosaic virus (type strain)	Tobamovirus
152.	Ribgrass mosaic virus	Tobamovirus
153.	Sunn-hemp mosaic virus	Tobamovirus
154.	Cucumber green mottle mosaic virus	Tobamovirus
155.	Odontoglossum ringspot virus	Tobamovirus
156.	Tomato mosaic virus	Tobamovirus
157.	Wheat yellow leaf virus	Closterovirus
158.	Onion yellow dwarf virus	Potyvirus
159.	Cherry rasp leaf virus	Nepovirus
160.	Myrobalan latent ringspot virus	Nepovirus
161.	Bidens mottle virus	Potyvirus
162.	Sweet potato mild mottle virus	Ipomovirus
163.	Strawberry crinkle virus	Cytorhabdovirus
164.	Citrus leaf rugose virus	Ilarvirus
165.	Raspberry bushy dwarf virus	Idaeovirus
166.	Narcissus tip necrosis virus	Carmovirus
167.	Wheat spindle streak mosaic virus	Bymovirus

168.	Desmodium yellow mottle virus	Tymovirus
169.	Oat necrotic mottle virus	Rymovirus
170.	Narcissus latent virus	Macluravirus
171.	Clitoria yellow vein virus	Tymovirus
172.	Rice necrosis mosaic virus	Bymovirus
173.	Cacao necrosis virus	Nepovirus
174.	Raspberry vein chlorosis virus	Rhabdoviridae
175.	Pangola stunt virus	Fijivirus
176.	Artichoke Italian latent virus	Nepovirus
177.	Panicum mosaic virus	Sobemovirus
178.	Cymbidium ringspot virus	Tombusvirus
179.	Soybean dwarf virus	Luteoviridae
180.	Brome mosaic virus	Bromovirus
181.	Red clover necrotic mosaic virus	Dianthovirus
182.	Carnation etched ring virus	Caulimovirus
183.	Orchid fleck virus	Rhabdoviridae
184.	Tobamovirus group	Tobamovirus genus
185.	Nepovirus group	Nepovirus genus
186.	Grapevine Bulgarian latent virus	Nepovirus
187.	Potato virus T	Trichovirus
188.	Rubus yellow net virus	Unassigned virus
189.	Nicotiana velutina mosaic virus	Unassigned virus
190.	Guinea grass mosaic virus	Potyvirus
191.	Dasheen mosaic virus	Potyvirus
192.	Bean golden mosaic virus	Begomovirus
193.	Kennedya yellow mosaic virus	Tymovirus
194.	Maize chlorotic dwarf virus	Waikavirus
195.	Daphne virus X	Potexvirus
196.	Frangipani mosaic virus	Tobamovirus
197.	Cowpea mosaic virus	Comovirus
198.	Raspberry ringspot virus	Nepovirus
199.	Comovirus group	Comovirus genus
200.	Potexvirus group	Potexvirus genus
201.	Lilac ring mottle virus	Ilarvirus
202.	Lilac chlorotic leafspot virus	Capillovirus
203.	Andean potato mottle virus	Comovirus
204.	Blueberry shoestring virus	Sobemovirus
205.	Sonchus yellow net virus	Nucleorhabdovirus
206.	Potato black ringspot virus	Nepovirus
207.	Beet yellow stunt virus	Closterovirus

208.	Satsuma dwarf virus	Nepovirus
209.	Cowpea severe mosaic virus	Comovirus
210.	Beet curly top virus	Curtovirus
211.	Alfalfa latent virus	Carlavirus
212.	Cowpea mottle virus	Carmovirus
213.	Cucumber mosaic virus	Cucumovirus
214.	Tymovirus group	Tymovirus genus
215.	Bromovirus group	Bromovirus genus
216.	Arracacha virus A	Nepovirus
217.	Oat sterile dwarf virus	Fijivirus
218.	Carrot thin leaf virus	Potyvirus
219.	Strawberry vein banding virus	Caulimovirus
220.	Maize rayado fino virus	Marafivirus
221.	Chloris striate mosaic virus	Mastrevirus
222.	Erysimum latent virus	Tymovirus
223.	Broad bean necrosis virus	Pomovirus
224.	Lucerne transient streak virus	Sobemovirus
225.	Lucerne Australian latent virus	Nepovirus
226.	Citrus exocortis viroid	Pospiviroid
227.	Hibiscus chlorotic ringspot virus	Carmovirus
228.	Heracleum latent virus	Trichovirus
229.	Alfalfa mosaic virus	Alfamovirus
230.	Turnip yellow mosaic virus	Tymovirus
231.	Bean mild mosaic virus	Carmovirus
232.	Tobacco leaf curl virus	Begomovirus
233.	Hibiscus latent ringspot virus	Nepovirus
234.	Tobacco necrotic dwarf virus	Luteoviridae
235.	Peanut clump virus	Pecluvirus
236.	Melandrium yellow fleck virus	Bromovirus
237.	Blackgram mottle virus	Carmovirus
238.	Quail pea mosaic virus	Comovirus
239	Maclura mosaic virus	Macluravirus
240.	Leek yellow stripe virus	Potyvirus
241.	Hop mosaic virus	Carlavirus
242.	Potato virus Y	Potyvirus
243.	Cauliflower mosaic virus	Caulimovirus
244.	Plant rhabdovirus group	Rhabdoviridae family
245.	Potyvirus group	Potyvirus genus
246.	Bean rugose mosaic virus	Comovirus
247.	Viola mottle virus	Potexvirus

248.	Rice ragged stunt virus	Oryzavirus
249.	Carrot red leaf virus	Luteoviridae
250.	Shallot latent virus	Carlavirus
251.	Cereal chlorotic mottle virus	Rhabdoviridae
252.	Galinsoga mosaic virus	Carmovirus
253.	Pepper mottle virus	Potyvirus
254.	Avocado sun-blotch viroid	Avsunviroid
255.	Peru tomato virus	Unassigned virus
256.	Tephrosia symptomless virus	Carmovirus
257.	Pea enation mosaic virus	Enamovirus
258.	Tobacco etch virus	Potyvirus
259.	Carlavirus group	Carlavirus genus
260.	Closterovirus group	Closterovirus genus
261.	Hop latent virus	Carlavirus
262.	American hop latent virus	Carlavirus
263.	Elderberry carlavirus	Carlavirus
264.	Foxtail mosaic virus	Potexvirus
265.	Helenium virus S	Carlavirus
266.	Plantain virus X	Potexvirus
267.	Blueberry leaf mottle virus	Nepovirus
268.	Beet leaf curl virus	Rhabdoviridae
269.	Rice stripe virus	Tenuivirus
270.	Arracacha virus B	Nepovirus
271.	Artichoke yellow ringspot virus	Nepovirus
272.	Pelargonium zonate spot virus	Unassigned virus
273.	Hypochoeris mosaic virus	Furovirus
274.	Southern bean mosaic virus	Sobemovirus
275.	Ilarvirus group	Ilarvirus genus
276.	Tulip virus X	Potexvirus
277.	Ullucus virus C	Comovirus
278.	Tobacco yellow dwarf virus	Mastrevirus
279.	Voandzeia necrotic mosaic virus	Tymovirus
280.	American plum line pattern virus	Ilarvirus
281.	Spinach latent virus	Ilarvirus
282.	Zucchini yellow mosaic virus	Potyvirus
283.	Maize white line mosaic virus	Unassigned virus
284.	Maize chlorotic mottle virus	Machlomovirus
285.	Artichoke vein banding virus	Nepovirus
286.	Bean leaf roll virus	Luteoviridae
287.	Coconut cadang-cadang viroid	Cocadviroid

288.	Asparagus virus 2	Ilarvirus
289.	Honeysuckle latent virus	Carlavirus
290.	Tomato ringspot virus	Nepovirus
291.	Potato leafroll virus	Polerovirus
292.	Papaya ringspot virus	Potyvirus
293.	Watermelon mosaic virus 2	Potyvirus
294.	Plant reovirus group	Reoviridae family
295.	Caulimovirus group	Caulimovirus genus
296.	Rice gall dwarf virus	Phytoreovirus
297.	African cassava mosaic virus	Begomovirus
298.	Radish yellow edge virus	Alphacryptovirus
299.	Rice hoja blanca virus	Tenuivirus
300.	Maize stripe virus	Tenuivirus
301.	Olive latent ringspot virus	Nepovirus
302.	Melon necrotic spot virus	Carmovirus
303.	Tomato golden mosaic virus	Begomovirus
304.	Wineberry latent virus	Potexvirus
305.	Blackeye cowpea mosaic virus	Potyvirus
306.	Cherry leaf roll virus	Nepovirus
307.	Tobacco streak virus	Ilarvirus
308.	Carnation ringspot virus	Dianthovirus
309.	Tobacco ringspot virus	Nepovirus
310.	Iris fulva mosaic virus	Potyvirus
311.	Poinsettia mosaic virus	Tymovirus
312.	Barley yellow striate mosaic virus	Cytorhabdovirus
313.	Tobacco stunt virus	Unassigned virus
314.	Yam mosaic virus	Potyvirus
315.	Carnation cryptic virus	Alphacryptovirus
316.	Potato virus V	Potyvirus
317.	Velvet tobacco mottle virus	Sobemovirus
318.	Solanum nodiflorum mottle virus	Sobemovirus
319.	Cucumber leaf spot virus	Carmovirus
320.	Rice grassy stunt virus	Tenuivirus
321.	Sweet clover necrotic mosaic virus	Dianthovirus
322.	Northern cereal mosaic virus	Cytorhabdovirus
323.	Mung bean yellow mosaic virus	Begomovirus
324.	Iris mild mosaic virus	Potyvirus
325.	Tobacco vein mottling virus	Potyvirus
326.	Hop stunt viroid	Hostuviroid
327.	Blueberry red ringspot virus	Caulimovirus

328.	Ginger chlorotic fleck virus	Sobemovirus
329.	Subterranean clover mottle virus	Sobemovirus
330.	Pepper mild mottle virus	Tobamovirus
331.	Soybean chlorotic mottle virus	Caulimovirus
332.	White clover cryptic virus 2	Betacryptovirus
333.	Black raspberry necrosis virus	Unassigned virus
334.	Cassia yellow blotch virus	Bromovirus
335.	Dioscorea latent virus	Potexvirus
336.	Nerine virus X	Potexvirus
337.	Bean common mosaic virus	Potyvirus
338.	Iris severe mosaic virus	Potyvirus
339.	Luteovirus group	Luteoviridae family
340.	Johnsongrass mosaic virus	Potyvirus
341.	Maize dwarf mosaic virus	Potyvirus
342.	Sugarcane mosaic virus	Potyvirus
343.	Lettuce necrotic yellows virus	Cytorhabdovirus
344.	Barley stripe mosaic virus	Hordeivirus
345.	Groundnut rosette assistor virus	Luteoviridae
346.	Tobacco rattle virus	Tobravirus
347.	Pepper ringspot virus	Tobravirus
348.	Miscanthus streak virus	Mastrevirus
349.	Apple scar skin viroid	Apscaviroid
350.	Pepino mosaic virus	Potexvirus
351.	Tobacco mild green mosaic virus	Tobamovirus
352.	Tombusvirus group	Tombusvirus genus
353.	Citrus tristeza virus	Closterovirus
354.	Potato virus X	Potexvirus
355.	Groundnut rosette virus	Umbravirus
356.	Barley mild mosaic virus	Bymovirus
357.	Cassava American latent virus	Nepovirus
358.	Squash leaf curl virus	Begomovirus
359.	Sorghum mosaic virus	Potyvirus
360.	Raspberry bushy dwarf virus	Idaeovirus
361.	Cassava Ivorian bacilliform virus	Unassigned
362.	Peach latent mosaic viroid	Avsunviroid
363.	Tospovirus genus	Tospovirus genus
364.	Peach rosette mosaic virus	Nepovirus
365.	Pear blister canker viroid	Apscaviroid
366	Potyviridae family	Potyviridae family
367	Narcissus late season yellow virus	Potyvirus

REFERENCES

Adams, A.N. (1978). The detection of plum pox virus in Prunus species in enzyme-linked immunosorbent assay (ELISA). Annals of Applied Biology, 90:215-221.

Adsur J. 1950. On the properties of sugarcane mosaic virus. Phytopathology 40: 214-215.

Angew. Chem. Int. Ed.1999, 38, 3269.

Atanassov, D. (1932). Plum pox. A new virus disease. Ann. Univ. Sofia, Fac. Agric. Silvic. 11, 49-69.

Barker, H., McGeachy, K.D., Ryabov, E.V., Commandeur, U., Mayo, M.A., and M. Taliansky (2001). Evidence for RNA-mediated defence effects on the accumulation of Potato leafroll virus. Journal of General Virology, 82, 3099-3106.

Barton T.J., L.M.Bull, W.G.Klemperer, D.A.Loy, B.McEnaney, M. Misono, P.A.Monson, G. Pez, G.W. Scherer, J.C.artuli, O.M. Yaghi, Chem. Mater.1999, 11, 2633.

Bawden F.C. 1959. Physiologyof virus diseases. Ann. Res. Plant Physiol 10: 239-253.

Bawden F.C. and B. Kassanis 1950. Some effects of plant nutrition on the multiplication of viruses. Ann. Appl. Biol. 37: 215-228.Bock K.R. 1982. Geminivirus diseases in tropical crops. Plant Disease 66:266-270.

Bawden, F.C., Pirie, N.W., Bernal, J.D. and Fankuchen, I., Nature (1936) 138, 1051-1053.

Beachy, R.N., Loesch-Fries, S., Tumer, N.E. (1990). Coat protein mediated resistance against virus infection. Ann. Rev. Phytopath. 28, 451-474.

Bernal, J.D. and Fankuchen, I., J. Gen.Physiol. (1941) 28, 111-165.

Block, J., A. Mackenzie, P. Guy, and A. Gibbs. 1987. Nucleotide sequence comparisons of turnip yellow mosaic virus isolates from Australia and Europe. Arch. Virol. 97:283-295.

Boscia, D., Zeramdini, H., Cambra, M., Potere, O., Gorris, M.T., Myrta, A., DiTerlizzi, B., and V. Savino. (1997). Production and characterization of a monoclonal antibody specific to the M serotype of plum pox potyvirus. European Journal of Plant Pathology, 102:477-480.

Boswell, K. and Gibbs, A. (1986). The VIDE data bank for plant viruses. In Developments and Applications in Virus Testing (ed. R.A.C. Jones). pp. 283-287. Assoc. Applied Biologists, U.K.

Boswell, K.F. and Gibbs, A.J. (eds) (1983). Viruses of Legumes 1983-Descriptions and Keys from VIDE. 139 pp. Australian National University, Canberra.

Boswell, K.F., Dallwitz, M.J., Gibbs, A.J. and Watson, L. (1986). The VIDE (Virus Identification Data Exchange) project: a database for plant viruses. Rev. Plant Path. 65: 221-231.

Brandes J. and B. Bercks 1969. Gross morphology and serology as a basis for clssification of elongate plant viruses. Pp 1-24. In : K.M. Smith and M.L. Lauffer (eds) Advances in virus research, Vol. 11. Academic Press, New York.

Brunt, A., Crabtree, K. and Gibbs, A. (1990). Viruses of Tropical Plants: Descriptions and Lists from the VIDE Database. 707 pp. C.A.B. International, U.K.

Brunt, A., Crabtree, K., Dallwitz, M., Gibbs, A. and Watson, L. (1996). Viruses of Plants: Descriptions and Lists from the VIDE Database. 1484 pp. C.A.B. International, U.K.

Büchen-Osmond, C., Crabtree, K., Gibbs, A. and McLean, G. (1988). Viruses of Plants in Australia. 590 pp. Australian National University, Canberra.

CABI/EPPO, (1998). Plum pox potyvirus. Distribution Maps of Quarantine Pests for Europe No. 320. Wallingford, UK, CAB International.

Calisher, C.H. and Fauquet, C.M. (1992). Steadman's ICTV Virus Words. 271 pp. Williams and Wilkins, Baltimore.

Cambra, M., Asensio, M., Gorris, M.T., Perez, E., Camarasa, E., Garcia, J.A., Moya, J.J., Lopez-Abella, D., Vela, C., and A. Sanz. (1994). Detection of plum pox potyvirus using monoclonal antibodies to structural and non-structural proteins. Bulletin OEPP/ EPPO, 24:569-577.

Canady M, Tsuruta H, Johnson J. 2001. Analysis of Rapid, Large-Scale Protein Quaternary Structural Changes: Time-Resolved X-ray Solution Scattering of Nudaurelia capensis w Virus (NwV) Maturation. J. Mol. Biol. 311:803-814

Canady MA, Tihova M, Hanzlik TN, Johnson JE, Yeager M. 2000. Large Conformational Changes in the Maturation of a Simple RNA Virus, Nudaurelia capensis omega Virus (NomegaV). J Mol Biol 299:573-584

Candresse, T., Cambra, M., Dallot, S., Lanneau, M., Asensio, M., Gorris, M.T., Revers, F., Macquaire, G., Olmos, A., Boscia, D., Quiot, J.B., and J. Dunez. (1998). Comparison of monoclonal antibodies and polymerase chain reaction assay for the typing of isolates belonging to the D and M serotypes of plum pox potyvirus. Phytopathology 88:198-204.

Capoor S.P. and P.M. Verma 1948. A mosaic disease of papaya in the Bombay Province. Curr. Sci. 17: 265-266.

Caspar, D.L.D., Nature (1956) 177, 928-928.

Cauliflower mosaic virus. J. gen.,Virol., 11: 129-138.

Chaudhury H.C. 1960. Spread of rugose mosaic and leaf roll in different varities of potato in the plains of West Bengal. Amer. Potato J. 37: 173-175.Francki R.I.B., R.G. Milne and T. Hatta 1985. Atlas of Plant viruses. Vols. I and II. Academic Press, London.

Cheetham A.K., C.R.Acad.Sci.Ser.IIc 1999, 2, 387.

Cheetham A.K., G.Fe ?ey,T.Loiseau, Angew. Chem.1999, 111, 3466;

Cheetham A.K., Angew. Chem.2001, 113, 2913; Angew. Chem. Int.

Chen, J., Torrance, L., Cowan, G.H., MacFarlane, S.A., Stubbs, G. and Wilson, T.M.A., Phytopathology (1997) 87, 295-301.

Christoff, A. (1934). Mosaikkrankheit oder Viruschlorose, bei Apfeln. Eine neue Viruskrankheit. Phytopath. Z. 7:521-536.

Chui S.S.Y., S.M.F.Lo, J.P.H. Charmant, A.G. Orpen, I.D.

Clark, M.F. and A.N. Adams. (1977). Characteristics of a microplate method of enzyme-linked immunosorbent assay for the detection of plant viruses. J. Gen. Virol. 34:475-483.

Conway J, Wikoff W, Cheng N, Duda R, Hendrix R, Johnson J, Steven A. 2001. Virus maturation via large subunit rotations and local refolding. Science 292:744-748

Crescenzi, A., Nuzzaci, M., Levy, L., Piazzolla, P., and A. Hadidi. (1994). Infezioni di sharka su ciliegio dolce in Italia meridionale. Informatore Agrario, 34:73-75.

Dallwitz, M. J., Paine, T. A., and Zurcher, E. J. (1993). User's Guide to the DELTA System: a general system for processing taxonomic descriptions. 4th edition. 136 pp. (CSIRO Division of Entomology: Canberra).

Dallwitz, M.J. (1980). A general system for coding taxonomic descriptions. Taxon 29: 41-46.

Damsteegt, V.D., Waterworth, H.E., Mink, G.I., Howell, W.E., and L. Levy. (1997). Prunus tomentosa as a diagnostic host for the detection of plum pox virus and other Prunus viruses. Plant Dis. 81:329-332.

Dolja, V.V., Boyko, V.P., Agranovsky, A.A. and Koonin, E.V., Virology (1991) 184, 79-86.

Domingo, E., and J. J. Holland. 1988. High error rates, population equilibrium and evolution of RNA replication systems. Pp. 3-36 in E. DOMINGO, J. J. HOLLAND, and P. AHLQUIST, eds. RNA genetics, vol. 3. CRC Press, Boca Raton, Fla.

Dopazo, J., F. Sobrino, E. L. Palma, E. Domingo, and A. Moya 1988. Gene encoding capsid protein VP1 of foot-and-mouth disease virus: a quasispecies model of molecular evolution. Proc. Natl. Acad. Sci. USA 85:68 1 l-68 15.

EPPO, (1999). EPPO PQR database (Version 3.8). Paris, France:EPPO.

Erickson, J.W. and Bancroft, J.B., Virology (1978) 90, 36-46. Goodman, R., Virology (1975) 68, 287-298.

Esau K. 1967. Anatomy of plant virus infection. Ann. Rev. Phyto path., 5:45-76.

Eugene W. Nester et al., Microbiology: A Human Perspective (Boston: McGraw-Hill, 1998), 16.

Fenner F. 1986. Classification and nomenclature of viruses. Classification and nomenclature of viruses (XIth Report of International Committee on Taxonomy of Viruses).

Fenner, F. (1976). Classification and Nomenclature of Viruses; 2nd Report of the International Committee on Taxonomy of Viruses. Intervirology 7: 1-116.

Finch, J.T., J. Mol. Biol. (1965) 12, 612-619.

Fitch, W. M., J. M. E. Leiter, X. LI, and P. Palese. 199 1. Positive darwinian evolution in human influenza A viruses, Proc. Natl. Acad. Sci. USA 88:4270-4274.

Forster P.M., P.M.Thomas,A.K.Cheetham,Chem.Mater.2002, in press.

Fraenkel-Conrat H. 1974. Descriptive catalogue of viruses. In: Comprehensive Virology. Vol.I Plenum Press, New York.

Franco-Lara, L.F., McGeachy, K.D., Commandeur, U., Martin, R.R., Mayo M.A. & Barker, H. (1999) Transformation of tobacco and potato with cDNA encoding the full-length genome of Potato leafroll virus: evidence for a novel virus distribution and host effects on virus multiplication. Journal of General Virology, 80, 2813-2822

Franklin R.E., Nature (1956) 177, 928-930.

Fulton R.W. 1974. The biological activity of heterogenous particle types of plant viruses. In: Viruses, Evolution and Cancer., pp 723-756. E. Kurstak and K. Maramorosch (Eds.) Academic Press, New York, 813pp.

Ganguly B., Pushkarnath and G.C. Upreti 1963. Effect of nutrition on plant growth and concentration of potato virus X and Y. Indian Potato J. 5: 44-47.

Gibbs A.J. 1969. Plant virus classification. Advan. Virus Res., 14: 263-328.

Gibbs A.J. and B.D. Harrison 1976. Plant virology-The principles. Edward Arnold, London. Pp. 292.

Gillespie, J. H. 1992. The causes of molecular evolution. Oxford University Press, Oxford.

Gojobori, T., E. N. Moriyama, and M. Kimura. 1990. Molecular clock of viral evolution and the neutral theory. Proc. Natl. Acad. Sci. USA 87: 10015-10018.

Gonzalez, A., Nave, C. and Marvin, D., Acta Cryst. (1995) D51, 792-804.

Gottwald, T.R., Avinent, L., Llacer, G., Mendoza, A.H.de, and M. Cambra. (1995). Analysis of the spacial spread of sharka (plum pox virus) in apricot and peach orchards in eastern Spain. Plant Disease, 79:266-277.

Goulden, M.G., Davies, J.W., Wood, K.R.and Lomonossoff, G.P., J. Mol. Biol. (1992) 227, 1-8.

Gregory, J. and Holmes, K.C., J. Mol. Biol.(1965) 13, 796-801.

Hammond, J., Puhringer, H., da Camara Machado, A., and M. Laimer da Camara Machado. (1998). A broad-spectrum PCR assay combined with RFLP analysis for detection and differentiation of plum pox isolates. Acta Hort. 472: 483-490.

Hartmann, W. (1998). Hypersensitivity -a possibility for breeding sharka resistant plum hybrids. Acta Hort. 472, 429-432.

Hayati J. and J.P. Verma 1985. Effect of some chemicals on tomato leaf curl virus infection of tomato. Indian J. Virol. 1(2): 152-156.

Hennig, B., and H. G. Wittmann. 1972. Tobacco mosaic virus: mutants and strains. Pp. 546-456.

Hirth L. 1976. Reconstitution of helical viruses other than TMV. Comp. Virology. 6: 39-63.

Holmes F.O. 1939. Handbook of phytopathogenic viruses. Burgess, Minneapolis., pp.221.

Howarth, A.J. and Vandemark, G.J. (1989). Phylogeny of geminiviruses. J. gen. Virol. 70: 2717-2727.

Hung P.P. 1976. Assembly of spherical plant and bacterial viruses., Comp. Virology. 6; 65-102.

J. Kado and H. D. Agrawal, eds. Principles and techniques in plant virology. Van Nostrand, New York.

J.Do,A.J.Jacobson,Inorg.Chem.2001, 40, 2468.

John V.T. 1957. A review of plant viruses in India. J. Madras Univ. 27B: 373-459.

Johnson J, Speir J. 1997. Quasi-equivalent viruses: A Paradigm for Protein Assemblies. J. Mol. Biol. 269:665-675.

Johnson J, Speir J. 1999. Principles of Virus Structure. In Encyclopedia of Virology, 2nd Edition, ed. Granoff, R. Webster, Academic Press Ltd., London pp. 1946-1956

Johnson J. 1927. The classification of plant viruses. Wisc. Univ. Agr., Exp., Sta., Res., Bull., 76: 1-16.

K. Saunders et al., "The earliest recorded plant virus disease," Nature, 422:831, April 24, 2003.

K. Maeda, Y.Kiyozumi, F.Mizukami, Angew. Chem.1994, 106, 2429; Angew.Chem.Int.Ed.Engl.1994, 33, 2335.

Kalashyan, Y.A., Bilkey, N.D., Verderevskaya, T.D., and E.V. Rubina. (1994). Plum pox potyvirus on sour cherry in Moldova. Bulletin OEPP/EPPO 24:645-649.

Kaplan, N. L., R. R. Hudson, and C. H. Langley. 1989. The "hitchhiking effect" revisited. Genetics 123:887-899.

Kegler, H., and W. Hartman. (1998). Present status of controlling conventional strains of plum pox virus. In: Hadidi, A., Khetarpal, R.K., and H. Koganezawa, eds. Plant Virus Disease Control. Pp. 616-628. APS Press, St. Paul, MN.

Kegler, H., Fuchs, E., Grüntzig, M., Schwarz, S. (1998). Some results of 50 years of research on the resistance to plum pox virus. Acta Virol. 42, 200-215.

Kerlan, C., and J. Dunez. (1979). Biological and serological differentiation of strains of sharka virus. Annales de Phytopathologie, 11:241-250.

Kim Y., D.Y.Jung,Bull.Korean Chem.Soc.2000, 21, 656.

Kim Y., E.W.Lee,D.P.Jung,Chem.Mater.2001, in press.

Kim Y.J., D.Y.Jung,Bull.Korean Chem.Soc.1999, 20, 827.

Kimura, M. 1983. The neutral theory of molecular evolution. Cambridge University Press, Cambridge.

Kreitman, M. 199 1. Detecting selection at the level of DNA. Pp. 204-221 In: R. K. Selander, A. G. Clark, and T. S. Whittam, eds. Evolution at the molecular level. Sinauer, Sunderland, Mass.

Lartey, R. T., Voss, T. C., & Melcher, U. (1996). Tobamovirus evolution: gene overlaps, recombination, and taxonomic implications. Mol. Biol. Evol., 13 (10), 1327-1338.

Lata R, Conway JF, Cheng N, Duda RL, Hendrix RW, Wikoff WR, Johnson JE, Tsuruta H, Steven AC. 2000. Maturation dynamics of a viral capsid: visualization of transitional intermediate states. Cell 100:253-63

Levy, L., and A Hadidi, (1994). A simple and rapid method for processing tissue infected with plum pox potyvirus for use with a specific 3' non-coding region RT-PCR assay. Bulletin OEEP/ EPPO 24:595-604.

Livage C., C.Egger,G.Fe ?ey,Chem.Mater.1999, 11, 1546.

Livage C., C.Egger,G.Fe ?ey,Chem.Mater.2001, 13, 410.

Llácer, G. and Cambra, M. (1998). Thirteen years of sharka disease in Valencia, Spain. Acta Hort. 472, 379-384.

Loff A., R.W. Hornee and P. Tournier 1962. A system of viruses. Cold Spring Harbor Symp. Quant. Biol., 27: 51-55.

Lopez-Moya, J.J., Fernandez-Fernandez, M.R., Cambra, M., and J.A. Garcia. (1999). Biotechnological aspects of plum pox virus. J. of Biotechnology 1999.

Luria S.E. 1959 Viruses as infective genetic material. In: V.Najjar (ed) Immunity and Virus Infection. John Wiley.

LYNCH, M., and T. J. CREASE. 1990. The analysis of population survey data on DNA sequence variation. Mol. Biol. Evol. 7:377-394.

Makowski, L., J. Appl. Cryst. (1978) 11, 273-283.

Malinowski, T., Zawadzka, B., Ravelonandro, M., Scorza, R. (1998). Preliminary report on the apparent breaking of resistance of a transgenic plum by chip bud inoculation of plum pox virus PPV-S. Acta Virol. 42, 241-243.

Mandahar C.L. 1978. Introduction to plant viruses., S. Chand and Co. Ltd., New Delhi. 333pp.

Maruyama, T., and C. W. Birky. 199 1. Effects of periodic selection on gene diversity in organelle genomes and other systems without recombination. Genetics 127:449-45 1.

Matthews, R. E. F. 1991. Plant virology. Academic Press, New York.

Maynard Smith, J. 1991. The population genetics of bacteria. Proc. R. Sot. Lond. [Biol.] 245:37-4 1.

Mayo, M.A., Ryabov, E., Fraser, G. & Taliansky, M. (2000) Mechanical transmission of Potato leafroll virus. J. Gen. Virol. 81, 2791-2795.

Mayo, M.A., Ryabov, E., Fraser, G. & Taliansky, M. (2000) Mechanical transmission of Potato leafroll virus. J. Gen. Virol. 81, 2791-2795.

Mehrotra R.S. 1988. Plant Pathology. In: Tata McGraw-hill Publishing Company Limited, New Delhi.

Meshi, T., M. ishikawa, N. Takamatsu, T. Ohno, and Y. Okada. 1983. The 5'-terminal sequence of TMV RNA: question on the polymorphism found in vulgare strains. FEBS Lett. 162:282-285.

Millane, R.P., Acta Cryst. (1989) A45, 573-576.

Murphy, F.A., Fauquet, C.M., Mayo, M.A., Jarvis, A.W., Ghabrial, S.A., Summers, M.D., Martelli, G.P. and Bishop, D.H.L. (1995). 6th Report of the International Committee on Taxonomy of Virusus. Archiv. Virol. Suppl. 10, Springer-Verlag, Wein, New York.

Myrta, A., Potere, O., Boscia, D., Candresse, T., Cambra, M., and Savino, V. (1998). Production of a Monoclonal specific to the El Amar Strain of Plum Pox Virus. Acta Virologica 42: 248-250.

Nagaich B.b. and K.S. Vashisth 1963. Barley yellow dwarf, a new virus disease for India. Indian Phytopath., 16: 318-319.

Namba, K. and Stubbs, G., Acta Cryst.(1987) A43, 533-539.

Namba, K., Pattanayek, R. and Stubbs, G.J. Mol. Biol. (1989) 208, 307-325.

Nambudripad, R., Stark, W. and Makowski, L., J. Mol. Biol. (1991) 220, 359-379.

NEI, M. 1987. Molecular evolutionary genetics. Columbia University Press, New York.

NEI, M., and F. TAJIMA. 1983. DNA polymorphism detectable by restriction endonucleases. Genetics 97: 145-163.

NEI, M., and J. C. MILLER. 1991. A simple method for estimating average number of nucleotide substitutions within and between populations from restriction data. Genetics 125:873-879.

Nemchinov, L., Crescenzi, A., Hadidi, A., Piazzolla, P., and T. Verderevskaya. (1998). Present status of the new cherry subgroup of plum pox virus (PPV-C). In: Hadidi, A., Khetarpal, R.K., and H. Koganezawa, eds. Plant Virus Disease Control. Pp. 629-638. APS Press, St. Paul, MN.

Nemeth, M. (1986). Virus, mycoplasma, and rickettsia diseases of fruit trees. Martinus Nijhoff Pub., Dordrecht.

Nemeth, M., and M. Kolber. (1983). Additional evidence on seed transmission of plum pox virus in apricot, peach, and plum, proved by ELISA. Acta Hort., 130:293-300.

Nurkiyanova K. M., Ryabov E.V., Commander U.,. Duncan, G. H., Canto, T., Gray, S.M., Mayo, M.A. and Taliansky M.E. (2000). Tagging Potato leafroll virus with the jellyfish green fluorescent protein gene. Journal of General Virology, 81, 617-626.

Okada Y. and T. Ohno 1976. Plant viruses with mixed phenotypes. Comp. Virology. 6; 103-129.

Olmos, A., Dasi, M.A., Candresse, T., and M. Cambra. (1996). Print-capture PCR: A simple and highly sensitive method for the detection of plum pox virus (PPV) in plant tissues. Nucleic Acids Research, 24:2192-2193.

Onuki M., K. Hanada, "Genomic structure of a geminivirus in the genus begomovirus from yellow vein-affected Eupatorium makinoi," Journal of General Plant Pathology, 66:176-181, 2000.

Padidam, M. Beachy, R.N. and Fauquet, C.M. (1995). Classification and identification of geminiviruses using sequence comparisons. J. gen. Virol. 76: 249-263.

Powar C.B. and H.F. Daginawala 1992. General Microbiology. In: Himalaya Publishing House, Delhi. Comp. Virology. 6; 1-37.

Ravelonandro, M., Dunez, J., Scorza, R., Labonne, G. (1998). Challenging transgenic plums expressing potyvirus coat protein genes with viruliferous aphids. Acta Hort. 472, 413-420.

Ravelonandro, M., Scorza, R., Bachelier, J.C., Labonne, G., Levy, L., Damsteegt, V., Callahan, A.M., and Dunez, J. (1997). Resistance of transgenic Prunus domestica to plum pox virus infection. Plant Dis. 81, 1231-1235.

Raychaudhuri S.P. and B. Ganguly 1968. A mosaic streak of wheat. Phytopath. Z. 62: 61-65.

Regenmortel M.H.V. van, C.M. Fauquet, D.H.L. Bishop, E.B. Carstens, M.K. Estes, S.M. Lemon, J. Maniloff, M.A. Mayo, D.J. McGeoch, C.R. Pringle, R.B. Wickner (eds.) Virus Taxonomy: Seventh Report of the International Committee on Taxonomy of Viruses. (Academic Press, San Diego, 2000).

Richards K.E. and R.C. Williams 1976. Assembly of tobacco mosaic virus invitro. Comp. Virology. 6; 1-37.

Robson R., J.Chem.Soc.Dalton Trans.2000, 3735.

Rodriguez-Cerezo, E., S. F. Elena, A. MOYA, and F. Garciaarenal. 1991. High genetic stability in natural populations of the plant RNA virus tobacco mild green mosaic virus. J. Mol. Evol. 32:328-332.

Russel G.J., E.A.C. Follett, J.H. Subak-Sharpe and B.D. Harrison 1971. The double stranded DNA of

Ryabov E.V., Robinson D.J., and Taliansky M.E. (1999). A plant virus-encoded protein facilitates long-distance movement of heterologous viral RNA Proceedings of the National Academy of Sciences of USA 96, 1212-1217.

Ryabov V.V., Oparka K.J., Santa Cruz S., Robinson D.J. and Taliansky M.E. (1998). Intracellular location of two groundnut rosette umbravirus proteins delivered by PVX and TMV vectors. Virology 242, 303-313.

Ryabov, E.V., Fraser, G., Mayo, M.A., Barker, H., and Taliansky, M.E. (2001). Umbravirus gene expression helps Potato leafroll virus to invade mesophyll tissues and to be transmitted mechanically between plants. Virology, 286, 363-372.

Ryabov, E.V., Roberts, I.M., Palukaitis, P., and Taliansky, M. E. (1999). Host-specific cell-to-cell and long-distance movements of cucumber mosaic virus are facilitated by the movement protein of groundnut rosette virus. Virology, 260, 98-108.

Santa Cruz, S., Chapman, S., Roberts, A.G., Roberts, I.M., Prior, D.A.M. and Oparka K.J., Proc. Natl. Acad. Sci. USA (1996) 93, 6286-6290.

Sastry K.S.M. and S.J. Sing 1974. Effect of yellow vein mosaic virus infection on growth and yield of okra crop. Indian Phytopath. 27: 294-297.

Scorza, R., Callahan, A.M., Levy, L., Damsteegt, V., Ravelonandro, M. (1998). Transferring potyvirus coat protein genes through hybridization of transgenic plants to produce plum pox virus resistant plums (Prunus domestica L.). Acta Hort. 472, 421-427.

Scorza, R., Ravelonandro, M., Callahan, A.M., Cordts, J.M., Fuchs, M., Dunez, J., and Gonsalves, D. (1994). Transgenic plums (Prunus domestica L.) express the plum pox virus coat protein gene. Plant Cell Repts. 14, 18-22.

Selander, R. K. 1985. Protein polymorphism and the genetic structure of natural populations of bacteria. Pp. 85-106 in T. OHTA and K. AOKI, eds. Population genetics and molecular evolution. Japan Scientific Societies, Tokyo, and Springer, Berlin.

Sherman, W.B., Lyrene, P.M. (1983). Handling seedling populations. p 66-73. IN: Moore, J.N. and Janick, J. (eds) Methods in fruit breeding. Purdue Univ. Press, West Lafayette, IN.

Siegel A. and V. Hariharasubramanian 1974. Reproduction of small plant RNA viruses. Comp. Virology. 2: 61-108.

Singh M.N. and S.M. Paul Khurana 1985. Aphid transmission chractiristics of potato virus Y. Indian Jour. Virol. 1: 42-48.

Singh R.S. 1989. Plant Diseases. In: Oxford and IBH Publishing CO. PVT. LTD., New Delhi. pp. 484-525.

Smith A.E. 1978. Cryptic initiation sites in eukaryotic virus mRNAs. FEBS. Symposium A2 (Copenhagen, 1977). 43: 37-46.

Smith K.M. 1957. A text book of plant virus diseases. Little Brown and Co., Boston.

Smith, D. B., and S. C. Inglis. 1987. The mutation rate and variability of eukaryotic viruses: an analytical review. J. Gen. Virol. 68:2729-2740.

Steinhauer, N. Depolo, and J. J. Holland. 1985. The quasispecies (extremely hetero-geneous) nature of viral RNA genome populations: biological relevance-a review. Gene 40: 1-8.

Stephen S. Morse, "Evolving Views of Viral Evolution: Towards an Evolutionary Biology of Viruses," History and Philosophy of the Life Sciences 14 (1992): 215.

Stubbs, G. and Diamond, R., Acta Cryst.(1975) A31, 709-718.

Stubbs, G. and Makowski, L., Acta Cryst.(1982) A38, 417-425.] Pattanayek, R. and Stubbs, G., J. Mol. Biol.(1992) 228, 516-528.] Stubbs, G., Namba, K. and Makowski, L.,Biophys. J. (1986) 49, 58-60.

Stubbs, G., Acta Cryst. (1989) A45, 254-258.

Syller J. 1980. Transmission of potato leaf roll virus by Myzus persicae from leaves and plants at different ages. Potato Research 23: 453-456.

Tollin, P., Bancroft, J.B., Richardson, J.F., Payne, N.C. and Beveridge, T.J., Virology (1979) 98, 108-115.

Tollin, P., Wilson, H.R., and Bancroft, J.B., J. Gen. Virol. (1980) 49, 407-410.

Ton van Helvoort, "History of Virus Research in the Twentieth Century: The Problem of Conceptual Continuity," History of Science 34 (1994): 185.

Uppal B.N., P.M. Verma and S.P. Capoor 1940. Yellow vein mosaic of bhindi. Curr. Sci. 9: 227-228.

Varveri, C., Candresse, T., Cugusi, M., Ravelonandro, M., and J. Dunez. (1988). Use of (32)P-labeled transcribed RNA probe for dot hybridization detection of plum pox virus. Phytopathology, 78:1280-1283.

Varveri, C., Ravelonandro, M., and J. Dunez. (1987). Construction and use of cloned cDNA probe for the detction of plum pox virus in plants. Phytopathology, 77:1221-1224.

Verhoeven, J.Th.J., de Haas, A.M., Roenhorst, J.W. (1998). Outbreak and eradication of plum pox potyvirus in the Netherlands. Acta Hort. 472, 407-409.

Wang, H. and Stubbs, G., Acta Cryst (1993) A49, 504-513.

Wang, H. and Stubbs, G., J. Mol. Biol.(1994) 239, 371-384.

Wang, H., Culver, J.N. and Stubbs, G., J.Mol. Biol. (1997) 269, 769-779.

Watson, L. and Dallwitz, M.J. (1991). The Families of Angiosperms: automated descriptions, with interactive identification and information retrieval. Aust. Sys. Bot. 4: 601-695.

Watson, L. and Dallwitz, M.J. (1992 onwards). 'The Families of Flowering Plants: Descriptions, Illustrations, Identification and Information Retrieval.

Watson, L., Dallwitz, M.J., Gibbs, A.J. and Pankhurst, R.J. (1988). Automated taxonomic descriptions. In Prospects in Systematics (ed D.L. Hawksworth). pp 292-304. Clarendon Press, Oxford.

Welsh, L.C., Symmons, M.F., Sturtevant, J.M., Marvin, D.A. and Perham, R.N., J.Mol. Biol. (1998) 283, 155-177.

Wetzel, T. Candresse, T., Macquaire, G., Ravelonandro, M., and J. Dunez. (1992). A highly sensitive immunocapture polymerase chain reaction method for plum pox potyvirus detection. J. Virological Methods, 39:27-37.

Wetzel, T. Candresse, T., Ravelonandro, M., and J. Dunez. (1991b). A polymerase chain reaction assay adapted to plum pox potyvirus detection. J. Virological Methods, 33:355-365.

Wetzel, T. Candresse, T., Ravelonandro, M., Delbos, R.P., Mazyad, H., Aboul-Ata, A.E., and J. Dunez. (1991a). Nucleotide sequence of the 3'-terminal region of the RNA of the El Amar strain of the plum pox potyvirus. J. of Gen. Virology, 72:1741-1746.

Wikoff W, Liljas L, Duda R, Tsuruta H, Hendrix R, Johnson J. 2000. Topologically linked protein rings in the bacteriophage HK97 capsid. Science 289:2129-2133.

Williams, Science 1999, 283, 1148.

Yamashita, I., Hasegawa, K., Suzuki, H., Vonderviszt, F., Mimori-Kiyosue, Y. and Namba, K., Nature Struct. Biol. (1998) 5, 125-132.

Yamashita, I., Suzuki, H. and Namba, K., J.Mol. Biol. (1998) 278, 609-615.

Yeraguntaiah R.C. and T.K. Nariani1963. Bean mosaic virus in India. Indian J. Micribiol., 3: 154-150.

Zheng L.M., T.Whitfield, X.Wang,A.J.Jacobson,Angew.Chem. 2000, 112, 4702; Angew. Chem.Int.Ed.2000, 39, 4528.

Zheng L.M., X.Q.Wang, Y.S.Wang,A.J.Jacobson,J.Mater.Chem. 2001, 11, 1100.

INDEX